The Pit and the Penauium

Standard Deviation and Curve Fitting

Teacher's Guide

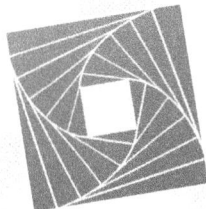

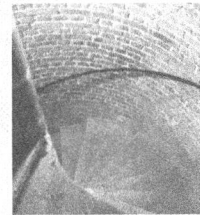

This material is based upon work supported by the National Science Foundation under award numbers ESI-9255262, ESI-0137805, and ESI-0627821. Any opinions, findings, and conclusions or recommendations expressed in this publication are those of the authors and do not necessarily reflect the views of the National Science Foundation.

Key Curriculum
1150 65th Street
Emeryville, California 94608
email: editorial@keypress.com
www.keycurriculum.com

First Edition Authors
Dan Fendel, Diane Resek, Lynne Alper, and Sherry Fraser

Contributors to the Second Edition
Sherry Fraser, Jean Klanica, Brian Lawler, Eric Robinson, Lew Romagnano, Rick Marks, Dan Brutlag, Alan Olds, Mike Bryant, Jeri P. Philbrick, Lori Green, Matt Bremer, Margaret DeArmond

Project Editors
Joan Lewis, Sharon Taylor

Consulting Editor
Mali Apple

Editorial Assistant
Juliana Tringali

Professional Reviewer
Rick Marks, Sonoma State University

Calculator Materials Editor
Christian Aviles-Scott

Math Checker
Christian Kearney

Production Director
Christine Osborne

Executive Editor
Josephine Noah

Textbook Product Manager
Timothy Pope

Publisher
Steven Rasmussen

Contents

Blackline Masters

What's Normal? Blackline Master
An (AB)Normal Rug Blackline Master
Standard Deviation Basics Blackline Master
Penny Weight Revisited Blackline Master
Can Your Calculator Pass This Soft Drink Test? Blackline Master
1-Inch Graph Paper Blackline Master
1/4-Inch Graph Paper Blackline Master
1-Centimeter Graph Paper Blackline Master
In-Class Assessment
Take-Home Assessment

Calculator Guide and Calculator Notes

Introduction

The Pit and the Pendulum Unit Overview

Intent

A classic short story, "The Pit and the Pendulum" by Edgar Allan Poe, is the context for this integrated mathematics and science unit. In the story, a prisoner is tied down while a pendulum with a sharp blade slowly descends above him. If the prisoner does not act, he will be killed by the swinging blade. At a point when the 30-foot pendulum seems to have about 12 swings left, the prisoner creates an escape plan. Given the facts of the story, students are asked to investigate this question: *Does the prisoner have time to execute his plan?*

To answer this question, students will have to decide what variables affect the time required for a pendulum to complete one swing. They will have to quantify the relevant relationships and use them to determine how long it would take for a 30-foot pendulum to swing 12 times.

Mathematics

This unit draws on and extends students' work in the first three units. It blends scientific experiments with the statistical concepts of normal distribution and standard deviation and the algebra of functions and graphs. The main concepts and skills that students will encounter and practice during the course of this unit are summarized below. References to graphing calculators should be understood to include other technology that might be available.

Experiments and Data

- Planning and performing controlled scientific experiments
- Working with the concept of period
- Recognizing and accommodating for the phenomenon of measurement variation
- Collecting and analyzing data
- Expressing experimental results and other data using frequency bar graphs

Statistics

- Recognizing the normal distribution as a model for certain kinds of data
- Making area estimates to understand the normal distribution
- Developing concepts of data spread, especially standard deviation
- Working with symmetry and concavity in connection with the normal distribution and standard deviation
- Applying standard deviation and the normal distribution in problem contexts

- Distinguishing between population standard deviation and sample standard deviation

- Calculating the mean and standard deviation of data sets, both by hand and with calculators

- Using standard deviation to decide whether a variation in experimental results is significant

Functions and Graphs

- Using function notation

- Using graphing calculators to explore the graphs of various functions

- Fitting a function to data using a graphing calculator

- Making predictions based on curve-fitting

Progression

The unit begins with an excerpt from "The Pit and the Pendulum." Students discuss whether the prisoner has time to execute his strategy, which leads to the statement of the unit problem. They then learn about measurement variation and the statistical ideas that help understand and account for it. Students then conduct a series of controlled experiments to determine the relevant variables and fit a curve to data to find a function they can use to answer the unit problem. There are 4 POWs in this unit.

Edgar Allan Poe—Master of Suspense

Statistics and the Pendulum

A Standard Pendulum

Graphs and Equations

Measuring and Predicting

Pacing Guides

50-Minute Pacing Guide (30 days)

Day	Activity	In-Class Time Estimate
Edgar Allen Poe—Master of Suspense		
1	The Pit and the Pendulum	15
	The Question	25
	POW 9: The Big Knight Switch	10
	Homework: Building a Pendulum	0
2	Discussion: Building a Pendulum	15
	Initial Experiments	30
	Homework: Close to the Law	5
3	Discussion: Close to the Law	15
	Discussion: Initial Experiments	30
	Homework: If I Could Do It Over Again	5
4	Discussion: If I Could Do It Over Again	5
	Time Is Relative	40
	Homework: What's Your Stride?	5
5	Discussion: What's Your Stride?	25
	Pulse Gathering	25
	Homework: Pulse Analysis	0
6	Presentations: POW 9: The Big Knight Switch	15
	Discussion: Pulse Analysis	20
	POW 10: Corey Camel	15
	Homework: Return to the Pit	0
7	Discussion: Return to the Pit	10
Statistics and the Pendulum		
	POW 10: Corey Camel (continued)	10
	Homework: What's Normal? (including introduction of normal distribution)	30
8	Discussion: What's Normal?	10
	A Mini-POW About Mini-Camel	35
	Homework: Flip, Flip	5
9	Discussion: Flip-Flip	20

	What's Rare?	25
	Homework: Penny Weight	5
10	Discussion: Penny Weight	20
	Mean School Data	30
	Homework: An (AB)Normal Rug	0
11	Discussion: An (AB)Normal Rug	10
	Data Spread	40
	Homework: Kai and Mai Spread Data	0
12	Discussion: Kai and Mai Spread Data	25
	Standard Deviation Basics	15
	Homework: The Best Spread	10
13	Discussion: The Best Spread (including discussion of the geometric interpretation of standard deviation)	15
	Making Friends with Standard Deviation	35
	Homework: Deviations	0
14	Discussion: Deviations	15
	Presentations: POW 10: Corey Camel	20
	Introduce: POW 11: Eight Bags of Gold	15
	Homework: Penny Weight Revisited	0
15	Discussion: Penny Weight Revisited (including discussion of standard deviation on the calculator)	45
	Homework: Can Your Calculator Pass This Soft Drink Test?	5
16	Discussion: Can Your Calculator Pass This Soft Drink Test?	20
A Standard Pendulum		
	The Standard Pendulum	30
	Homework: Standard Pendulum Data and Decisions	0
17	Discussion: Standard Pendulum Data and Decisions	10
	Pendulum Variations	35
	Homework: A Picture Is Worth a Thousand Words	5
18	Discussion: A Picture Is Worth a Thousand Words	10
	Pendulum Variations (continued)	30
	Homework: Pendulum Conclusions	10
19	Discussion: Pendulum Conclusions	15
	POW Revision	35

	Homework: POW Revision	0
20	Presentations: POW 11: Eight Bags of Gold	20
	Introduce: POW 12: Twelve Bags of Gold	30

Graphs and Equations

	Homework: Maliana the Market Analyst	0
21	Discussion: Maliana the Market Analyst	15
	Birdhouses	30
	Homework: So Little Data, So Many Rules	5
22	Discussion: So Little Data, So Many Rules	15
	Graphing Free-for-All	30
	Homework: Graphs in Search of Equations I	5
23	Discussion: Graphs in Search of Equations I	15
	Graphing Free-for-All (continued)	35
	Homework: Graphs in Search of Equations II	0
24	Discussion: Graphs in Search of Equations II	15
	Graphing Free-for-All (continued)	30
	Homework: Graphing Summary	5
25	Discussion: Graphing Summary	25

Measuring and Predicting

	An Important Function	25
	Homework: Graphs in Search of Equations III	0
26	Discussion: Graphs in Search of Equations III	10
	The Thirty-Foot Prediction	40
	Homework: Mathematics and Science	0
27	Discussion: Mathematics and Science	10
	The Thirty-Foot Prediction (continued)	35
	Homework: Beginning Portfolios	5
28	Presentations: POW 12: Twelve Bags of Gold	20
	Discussion: Beginning Portfolios	10
	The Pit and the Pendulum Portfolio	20
29	In-Class Assessment	40
	Homework: Take-Home Assessment	10
30	Assessment Discussion	30
	Unit Reflection	20

90-Minute Pacing Guide (18 days)

Day	Activity	In-Class Time Estimate
Edgar Allen Poe—Master of Suspense		
1	The Pit and the Pendulum	15
	The Question	25
	Initial Experiments	35
	POW 9: The Big Knight Switch	10
	Homework: Building a Pendulum and Close to the Law	5
2	Discussion: Building a Pendulum	10
	Discussion: Close to the Law	10
	Initial Experiments (continued)	25
	If I Could Do It Over Again	15
	Time Is Relative	30
	Homework: What's Your Stride?	0
3	Discussion: What's Your Stride?	25
	Pulse Gathering	25
	Pulse Analysis	40
	Homework: Return to the Pit	0
4	Discussion: Return to the Pit	15
	Presentations: POW 9: The Big Knight Switch	20
	POW 10: Corey Camel	20
Statistics and the Pendulum		
	Homework: What's Normal? (including introduction of normal distribution)	35
5	Discussion: What's Normal?	10
	A Mini-POW About Mini-Camel	35
	Flip, Flip	40
	Homework: Penny Weight	5

13	Presentations: POW 11: Eight Bags of Gold	20
	POW 12: Twelve Bags of Gold	30
	Birdhouses	30
	Homework: So Little Data, So Many Rules	10
14	Discussion: So Little Data, So Many Rules	10
	Graphing Free-for-All	30
	Graphs in Search of Equations I	20
	Graphing Free-for-All (continued)	30
	Homework: Graphs in Search of Equations II	0
15	Discussion: Graphs in Search of Equations II	10
	Graphing Free-for-All (continued)	20
	Graphing Summary	40
Measuring and Predicting		
	An Important Function	20
	Homework: Graphs in Search of Equations III	0
16	Discussion: Graphs in Search of Equations III	10
	The Thirty-Foot Prediction	75
	Homework: Mathematics and Science and Beginning Portfolios	5
17	Discussion: Mathematics and Science	10
	Discussion: Beginning Portfolios	10
	Presentations: POW 12: Twelve Bags of Gold	30
	The Pit and the Pendulum Portfolio	30
	Homework: Take-Home Exam	10
18	In-Class Assessment	40
	Assessment Discussion	30
	Unit Reflection	20

Materials and Supplies

All IMP classrooms should have a set of standard supplies and equipment. Students are expected to have materials available for working at home on assignments and at school for classroom work. Lists of these standard supplies are included in the section "Materials and Supplies for the IMP Classroom" in *A Guide to IMP*. There is also a comprehensive list of materials for all units in Year 1.

Listed below are the supplies needed for this unit. General and activity-specific blackline masters are available for presentations on the overhead projector or for student worksheets.

The Pit and the Pendulum

- String (non-stretch); unwaxed dental floss or fishing line works well
- Paper clips and masking tape
- Meter sticks or tape measures (one or two per group)
- Poster-size grid paper
- Washers (about 10 per group)
- Stop watches (one or two per group)
- Protractors (one or two per group)
- Scissors (one or two per group)
- Beans or other counters (at least 45 per pair of students)
- Coins (10 per student)
- Materials to make a 30-foot pendulum

More About Supplies

- The pendulum experiment supplies (string, tape, paper clips, tape measures, washers, stop watches, protractors, and scissors) can be kept in a basket or other container—one for each group. Many teachers have found that dental floss is a good substitute for string. Only one package of floss is needed in each group basket.

- Graph paper is a standard supply for IMP classrooms. Blackline masters of 1-Centimeter Graph Paper, 1/4-Inch Graph Paper, and 1-Inch Graph Paper are provided so that you can make copies and transparencies for your classroom. (You'll find links to these masters in "Materials and Supplies for Year 1" of the Year 1 guide and in the Unit Resources for each unit.)

Assessing Progress

The Pit and the Pendulum concludes with two formal unit assessments. In addition, there are many opportunities for more informal, ongoing assessment throughout the unit. For more information about assessment and grading, including general information about the end-of-unit assessments and how to use them, consult "Assessment and Grading" in the *A Guide to IMP*.

End-of-Unit Assessments

This unit concludes with in-class and take-home assessments. The in-class assessment is intentionally short so that time pressures will not affect student performance. Students may use graphing calculators and their notes from previous work when they take the assessments.

Ongoing Assessment

Assessment is a component in providing the best possible ongoing instructional program for students. Ongoing assessment includes the daily work of determining how well students understand key ideas and what level of achievement they have attained in acquiring key skills.

Students' written and oral work provides many opportunities for teachers to gather this information. Here are some recommendations of written assignments and oral presentations to monitor especially carefully that will offer insight into student progress.

- *Initial Experiments:* This activity will tell you how well students understand the idea of isolating a single variable.

- *Pulse Analysis:* This assignment will tell you about students' understanding of mean and frequency bar graphs.

- *Kai and Mai Spread Data:* This activity will give you a baseline of information about students' understanding of data spread.

- *Penny Weight Revisited:* This activity will guide you in determining students' intuitive understanding of standard deviation.

- *Pendulum Conclusions:* This assignment will tell you how well students can reason using the concept of standard deviation.

- *Graphing Summary:* This activity will give you information on what students know about the shape of graphs of various functions.

- *Mathematics and Science:* This assignment will give you insight into what students see as the key ideas of the unit.

Supplemental Activities Overview

The Pit and the Pendulum contains a variety of activities at the end of the student pages that you can use to supplement the regular unit material. These activities fall roughly into two categories.

- **Reinforcements** increase students' understanding of and comfort with concepts, techniques, and methods that are discussed in class and are central to the unit.

- **Extensions** allow students to explore ideas beyond those presented in the unit, including generalizations and abstractions of ideas.

The supplemental activities are presented in the teacher's guide and the student book in the approximate sequence in which you might use them. Below are specific recommendations about how each activity might work within the unit. You may wish to use some of these activities, especially the later ones, after the unit is completed.

Poe and "The Pit and the Pendulum" (extension) This research assignment—which asks students to learn more about the story and the author, Edgar Allan Poe, and report on their findings—can be given to students at any point in the unit, perhaps right after reading the excerpt from the story. You may want to inform students' English teachers about this activity and develop some form of collaboration.

Getting in Synch (extension) This activity presents the concept of *period* in a different context, through which students will explore some number theory. The activity can be assigned at any time, as the term period is introduced in the first activity. For whole-number periods, the questions are fairly straightforward, but for fractions, the issues get more interesting. Question 3 hints at an idea that Greek mathematicians called *incommensurate numbers*. This refers to a pair of numbers that we would today describe as having a ratio that is an **irrational number**. Don't expect students to get the correct answer to this question.

Height and Weight (reinforcement or extension) The emphasis in this activity is on the general process of investigation. It asks students to explore whether the weight of something is correlated with its height (in a context of the students' own choosing). This activity relates to the discussion in *Initial Experiments* about what variables are related to the period of a pendulum's swing.

Octane Variation (reinforcement) This activity asks students to assess the validity of potential conclusions based on a small data set and to propose improvements to the experiment. It can be used at any time after the discussion of *Close to the Law*.

Stem-and-Leaf Plots and Quartiles and Box Plots These reference materials summarize two methods of representing numeric data and introduce the concept of quartiles, an extension of the concept of median, which is used in creating box-and-whiskers plots.

Data Pictures (reinforcement) This activity provides an opportunity for students to construct three different types of visual representations of a data set—a frequency bar graph, a stem-and-leaf plot, and a box plot—and examine the advantages and disadvantages of each. This activity can be used in conjunction with the reference pages *Stem-and-Leaf Plots* and *Quartiles and Box Plots*.

More Knights Switching (extension) This activity is similar to *POW 9: The Big Knight Switch* and is a good follow-up to the discussion of that activity.

A Knight Goes Traveling (extension) This activity is similar to the other knight-switch problems but uses a square chessboard of some unknown size. This is another good follow-up investigation to *POW 9: The Big Knight Switch*.

Data for Kai and Mai (reinforcement) This activity poses questions similar to Question 4 of *The Best Spread*.

Making Better Friends (extension) This activity provides additional opportunities for students to explore standard deviation, continuing the challenges posed in *Making Friends with Standard Deviation*.

Mean Standard Dice (reinforcement or extension) This activity gives students practice with standard deviation. It also demonstrates that doing the same experiment more times doesn't change either the mean or the standard deviation. Students may be surprised about the result for standard deviation in Question 3.

More About Soft Drinks, Calculators, and Tests (reinforcement or extension) This activity follows up on *Can Your Calculator Pass This Soft Drink Test?* through questions about probabilities related to the normal curve. The activity involves an asymmetric "tail" problem (beyond two standard deviations at one end, beyond one standard deviation at the other end). Questions 2 and 3 require students to estimate areas under the normal curve whose boundaries are not whole multiples of the standard deviation from the mean.

Are You Ambidextrous? (reinforcement or extension) This experiment-based activity involves the idea of standard deviation and reasoning similar to that needed in the main unit problem. You may want to use this activity either as preparation for *Pendulum Variations* or as a follow-up to that activity.

Family of Curves (reinforcement or extension) This activity can be used with groups to extend or reinforce the ideas in Graphing Free-for-All. Give each group the equation of a basic curve, such as $y = x^2$, and ask them to look at some simple changes that could be made to the equation, such as $y = x^2 + 2$ or $y = 3x^2$. Groups should then explore how graphs vary among members of that family of curves and make a poster showing their results. They can use graphing calculators or computers to assist with their work.

More Height and Weight (extension) This activity follows up on the supplemental activity *Height and Weight* by posing the question with a more specific focus. It also incorporates students' work with curve-fitting.

Out of Action (reinforcement or extension) In this activity, students will fit a curve to data to make a prediction and then use that prediction to make a decision. This activity might be used late in the unit, as a follow-up to students' work on the unit problem.

Gettin' On Down to One In this POW-like activity, students investigate the behavior of number sequences and the relationships between that behavior and the number chosen to start a particular sequence.

Programming Down to One In this activity, students write a computer or calculator program to generate the number sequences in *Gettin' on Down to One.*

The Pit and the Pendulum: A Mathematical Commentary

Victor J. Donnay, Professor of Mathematics, Bryn Mawr College,
Bryn Mawr, Pennsylvania

Edward F. Wolff, Chair, Computer Science and Mathematics,
Arcadia University, Glenside, Pennsylvania

Is the Interactive Mathematics Program (IMP) unit *The Pit and the Pendulum* a literature unit, exploring the gripping narrative techniques that Edgar Allen Poe employs in his famous story? Is it a physics unit, experimentally investigating the harmonic motion of pendulums? One can understand the confusion a first-time reader of the unit might have. It is only as the unit progresses that the main focus becomes clear: the unit is aimed at having students experiment with, discover, create, and understand the main principles and techniques of statistical inference and mathematical modeling and, in so doing, experience for themselves the process of mathematical investigation. Along the way, students have the opportunity to connect their new experiences to their everyday understanding of the world and to their previous mathematical knowledge in ways that reflect the most up-to-date understanding of how people learn.

The unit is launched with Poe's story and the challenge, "Does the story's hero really have time to carry out his escape plan?" To determine the answer, students build their own pendulum and investigate the time it takes for the pendulum to swing back and forth (its period). There are many factors that could potentially influence the period of a pendulum (e.g., initial amplitude, length of the pendulum, mass of the bob) and students will need to investigate these multiple factors on the way to answering the unit question. Finally, with stopwatches in hand, students get down to the nitty-gritty of measuring the period of their pendulum. However, even with this simple task, the authors add complexity by making students repeat their measurements multiple times and focusing student attention on the variation in their measurements.

Clearly the authors have a point of view: mathematics can be used to address interesting (and perhaps important) real-world problems, but since the real world is fraught with complexity, our mathematics needs to be able make sense of this complexity. Those who think that math problems have only one answer, that this answer is determined via a standard algorithm, and that the purpose of math class is to train students in successfully applying the algorithm are initially likely to be surprised and even frustrated by the authors' approach. Is this really mathematics?

In their measurement of periods, students grapple with the question of how much variation one would expect to find in measurements simply from the inherent experimental error (referred to by statisticians as sampling error) of repeating a measurement. When is the variation so large that some other factor besides experimental error is probably involved? Students grapple with this question as they decide for themselves what constitutes normal and abnormal variations. While as mathematicians, we are used to the process of introducing new terms and definitions, trying them out, and then deciding if they are worth keeping, many nonmathematicians have the misconception that mathematical definitions are absolute, not something for discussion or debate. By leading students through the process of testing out several alternative ways of characterizing data spread, the authors give students opportunities to experience for themselves this process of mathematical investigation, creation, and refinement. Students see that complex mathematical ideas can grow out of common-sense experience and that they have the ability to engage in such creative processes themselves. At the end of the process, students have been led to the sophisticated notion of standard deviation as a measure of variability. Students then gain experience with this new definition by "making up" data sets that display a range of standard deviations. This is a very important mathematical practice: exploring the logical implications of a definition by seeing what it implies and does not imply via examples and counterexamples.

Of course, the authors' inquiry approach to this topic takes longer than if the definition of standard deviation were simply given on the first day of class and students were then drilled in how to apply the formula. But as the research into how people learn has shown (see, for example, *How Students Learn: History, Mathematics, and Science in the Classroom,* National Research Council, 2005), and that has been confirmed for many of us through our own teaching experiences, students who are trained via such a direct-lecture approach, although perhaps able to perform successfully in the short term on a test, will quickly forget the information and not be able to transfer the concepts to new and novel situations. Some years later, such students may even claim ignorance of ever having heard of standard deviation, much less remember what it was about. If, as a result of their hands-on inquiry experiences, students are able to remember the main ideas and how and when to use them, those of us who teach them in college will be very grateful to their secondary teachers for this extra time investment.

Once students become comfortable with the concept of standard deviation, they immediately put it to good use. They look at several data sets, all of which they generated themselves, and come to the conclusion that for normally distributed data, results that lie more than two standard deviations from the expected mean are "rare" and thus, when encountered, lead one to suspect that the model being tested (what statisticians would call the null hypothesis) is not the one that governs the experimental data. This realization puts them in position to resolve the unit question. They first construct a standard pendulum, and take multiple measurements of its period to establish what the distribution of such measurements would look like when variation is due to measurement error alone. Next, they perform a series of controlled experiments in which they vary each of the three potential factors, one at a time, and plot the resulting periods on their standard pendulum distributions. They find that only a change in length leads to "rare" results and thus identify length as the factor that matters. Although there are also variations in the measured period when weight or mass change, these variations are small and hence can be attributed to measurement error.

Students are engaging in genuine statistical reasoning. To test a hypothesis such as "the length of the pendulum makes a difference in the period," statisticians would first establish what the distribution of periods would look like under the null hypothesis, which states that length makes no difference. This is what students are doing when they take and plot multiple measurements of the standard pendulum. Next, statisticians would gather data under the experimental condition (changing the length of the pendulum) and determine where those data lie on the null distribution they found earlier. If the data lie more than two standard deviations from the mean, students know that such results are "rare" and that the null distribution is likely *not* to be the one governing the data. That is, they reject the null hypothesis and conclude that length does impact period. A statistician would express this by saying that the data fall in the critical region and hence are statistically significant. All these concepts are covered in a standard elementary statistics course. The beauty of this unit is that students are guided to discover the concepts for themselves.

Having determined that it is the length of a pendulum that determines its period, students focus on determining exactly how the period varies as the length varies. At this point, what had been a statistical question transitions into a more traditional math-modeling problem: what kind of function can best be used to represent the data? Using calculators, students develop a tool chest of basic functions and their graphs and then determine which functional type best fits their data. From this perspective, students realize that functions and graphs are tools that one uses to understand data coming from real-world problems. The authors' approach is in contrast with the axiomatic approach in which functions and their properties are studied in a data- and context-free setting.

Ultimately, students determine the general form of the function (it turns out to be $p = c\sqrt{l}$, where p and l are the pendulum's period and length, respectively) and then experiment with the values of the function parameter c to find the specific function that best fits their pendulum data. They use the function to predict the period of Poe's 30-foot pendulum and then, in the unit's grand finale, construct such a pendulum to gauge the accuracy of their prediction.

By unit's end, students have experienced the essence of mathematical modeling: preexisting information is used to develop a potential model and make predictions. The model is then checked against data generated in new experiments and, if necessary, refined further. Such mathematical modeling is becoming every more prevalent in modern science. Thus the unit helps prepare students to be informed citizens, able to intelligently engage in societal decision making that will involve evaluating scientific and mathematical modeling.

The Problems of the Week in this unit provide students with rich opportunities to engage in mathematical thinking. Their emphasis on having students justify their results ("How do you know your answer is correct and is the best possible?") is a good way to lead students to precision and, ultimately, proofs. Unlike many secondary students who believe that proofs appear only in geometry courses, these students will gain an appreciation of their relevance throughout mathematics. One of the Problems of the Week, *Twelve Bags of Gold,* is especially difficult and, as the authors acknowledge, it is likely that very few students will be able to solve it. We think its inclusion is admirable. The challenge of persevering with a very difficult problem in order to gain even partial solutions is familiar to mathematicians, who struggle with open problems all of the time. For example, the recently proven

Fermat's Last Theorem was open for 300 years. However, along the way attempts to solve it led to new and valuable mathematics.

The Problems of the Week and their solutions remind us of the problem-solving heuristics developed by the well-known mathematician George Polya (see *How to Solve It*, George Polya, 1971). One such Polya strategy for solving a difficult problem is to start by trying to find the answer to a simpler version of the problem. The *A Mini-POW About Mini-Camel* activity, assigned a couple of days after the challenging *Corey Camel* POW, is a perfect example of this.

We now turn to a discussion of the why the sample variance is divided by $n - 1$. The authors introduce the sample standard deviation and state that it is a better way of estimating the true variance. For those interested, we provide some details regarding why statisticians do indeed prefer the quantity $s = \sqrt{\dfrac{\sum_{i=1}^{n}(x_i - \bar{x})^2}{n-1}}$ over

$$s_n = \sqrt{\dfrac{\sum_{i=1}^{n}(x_i - \bar{x})^2}{n}} \; .$$

As discussed in the IMP Year 4 unit *The Pollster's Dilemma*, s^2, the sample variance, is a so-called *unbiased* estimator for the population variance σ^2. That means that if one sampled over and over again, the values of s^2 that one would obtain from each sample would average out in the long run to equal σ^2, whereas averaging out s_n^2 would return an answer that is too small. Essentially, this is because numbers in a sample are always closer (in terms of squared distance) to their own mean than they are to any other number, including the overall population mean. This can be shown using elementary calculus.

Consider the quantity $\sum_{i=1}^{n}(x_i - a)^2$. Taking the derivative with respect to a and setting it equal to zero and solving for a will show that the value of a that minimizes the quantity is $\bar{x}$. This implies that any other value of a, including the true overall mean, would yield a larger answer. Consequently, in order to have an unbiased estimator, we need to increase s_n, which is accomplished by decreasing the denominator. Of course, this explanation does not explain why we divide by $n - 1$. A full, easy-to-follow proof (no knowledge of calculus is necessary) showing that $n - 1$ is indeed the correct denominator to use can be found online at http://blue.butler.edu/~orris/data/Nminus1.pdf.

Finally, it is interesting to note that even though s^2 is an unbiased estimator for σ^2, s itself is not an unbiased estimator for σ. (This may seem like a contradiction, but follows from the fact that if we average the squares of numbers and then take a square root, we do not obtain the average of the original numbers.) Finding an unbiased estimator for σ is a far more complicated problem, so we are happy to at least have a fairly simple unbiased estimator for σ^2.

One final thought: We have found that many students arrive at college with the sense that mathematics is just memorizing and applying arbitrary rules that are not good for anything useful. We feel that this unit debunks these misconceptions for students, showing them that math is a creative endeavor, developed by real people for understandable reasons and that they can engage in this process. Thus are mathematicians created.

Edgar Allan Poe—Master of Suspense

Intent

These activities use an excerpt from Edgar Allen Poe's short story "The Pit and the Pendulum" to pose the unit problem. Students conduct some initial exploration into the unit problem and then explore related activities that build a foundation for subsequent ideas in the unit.

Mathematics

To solve the unit problem, students have to identify the variable that affects the period of a pendulum, gather data using controlled experiments, find a function that fits these data, and then use the function to make a prediction. In *Edgar Allen Poe—Master of Suspense,* the focus is on identifying candidates for variables and learning important ideas about doing controlled experiments and measuring quantities. In particular, students will display data collected from a series of experiments and use these displays to come to understand the idea of measurement variation.

Progression

The unit begins with students reading excerpts from Poe's story, building a pendulum, and stating the unit problem. Then they do three different data-gathering and analyzing activities. In addition, students complete the first POW of the unit and start work on the second one.

The Pit and the Pendulum

The Question

POW 9: The Big Knight Switch

Building a Pendulum

Initial Experiments

Close to the Law

If I Could Do It Over Again

Time Is Relative

What's Your Stride?

Pulse Gathering

Pulse Analysis

POW 10: Corey Camel

Return to the Pit

The Pit and the Pendulum

Intent

The unit opens with an excerpt from Edgar Allan Poe's story "The Pit and the Pendulum." In the activity *The Question,* students will examine the story for clues related to whether the hero would have enough time to carry out his plan, thereby defining the unit problem.

Mathematics

The excerpt describes a situation in which a prisoner, about to be sliced by a blade attached to a pendulum swinging overhead, devises a plan to escape. This provides the setting for the question posed in the first activity of the unit: *Does the prisoner have time to execute his escape plan?*

Progression

Students will read along as the story is read and then discuss the key elements of the passage.

Approximate Time

15 minutes

Classroom Organization

Whole class

Materials

Recording of the excerpt (optional)

Doing the Activity

Read the excerpt to the class, have students take turns reading it aloud, or play a recording of the excerpt as students follow along. Don't worry if students do not understand every word; the important thing is that they get a general sense of what is going on. There is no need to define vocabulary before reading the excerpt, as students will derive meaning from the context.

You may want to mention that Edgar Allen Poe wrote "The Pit and the Pendulum" in 1842. The complete story is available from many online sources, such as the Edgar Allan Poe Society of Baltimore. The "pit" referred to in the title is the centerpiece of the first part of the full story.

Discussing and Debriefing the Activity

Ask students to pair up and spend a minute each describing their interpretations of the excerpt. If vocabulary becomes a concern, define words as necessary. Most of the uncommon terms will be clarified during the activity *The Question.*

Key Question

What are the key elements of this story?

Supplemental Activity

Poe and "The Pit and the Pendulum" (extension) asks students to learn more about the story and the author, Edgar Allan Poe, and report on their findings.

The Question

Intent

Students reexamine the excerpt from the short story "The Pit and the Pendulum," this time with more specific questions in mind.

Mathematics

The pendulum in the story is 30 feet long, and the prisoner estimates that it will swing 12 times from when he develops a plan of escape until he can carry out the plan. The goal in this activity is for students to state the mathematical question that will frame the unit: *Does the prisoner have time to execute his plan? What does one have to know about this situation to predict how long these 12 swings will take?*

Progression

Students revisit the excerpt to search for information they can use to answer the question of whether the story's hero really has time to carry out his escape plan. A class discussion will then help them to restate the question in more quantitative terms.

Approximate Time

25 minutes

Classroom Organization

Groups, followed by whole-class discussion

Doing the activity

Before students reexamine the excerpt in their groups, it may be helpful to have them generate a list of "things we know" and "things we need to know" to focus their thinking on the question of whether the prisoner can escape.

Discussing and Debriefing the Activity

As groups report their findings, list their responses in two categories: "What We Know" and "What We Need to Find Out."

Several specific pieces of information should be noted. Point them out if students do not see them.

- The ceiling is "some thirty or forty feet overhead" (first paragraph), so the pendulum is likely to be about this long.

- The pendulum is "within three inches of my bosom" (sixth paragraph) when the prisoner develops his plan.

- The pendulum will take "some ten or twelve vibrations" (seventh paragraph) before it reaches him.

- "Yet one minute" (near the end of the next-to-last paragraph) is all that is needed until the bandage would be loosened by the rats.

A more formal name for swing is period. Introduce this term as the amount of time it takes for a pendulum to make a complete swing (back *and* forth).

This discussion should also bring out that the story might not give students all the information they need (and not all the information in the story is necessarily needed). Explain that, in order to give the initial question some direction and specificity, students must make some assumptions. For example, since they are interested only in the last few swings of the pendulum, they might ignore the fact that the blade is moving down and that the length of the pendulum is changing.

Tell students that this unit will use 30 feet as the pendulum's length (from Poe's "some thirty or forty feet overhead") and 12 swings as the duration (from "some ten or twelve vibrations"). However, other assumptions may be needed later. For now, students will explore the following narrower question:

How long would it take for Poe's pendulum to make 12 swings?

Post this question on the wall. Students will be working on it, with some digressions, for the rest of the unit.

Tell the class that the basic goal of the unit will be to answer this revised question. Ask, What information will you need to answer this new question? How might you get that information? What materials will you need?

If students suggest that they could construct a pendulum like Poe's, tell them that they will eventually do so. However, ask whether they really have enough information at this point to construct a full-scale model of the pendulum. In particular, they don't yet know which facts about the pendulum are important. You might also bring out that in the real world, actual construction before creating a mathematical model is sometimes too expensive, impractical, or even impossible.

Explain that the unit focuses on trying to devise an indirect method to answer the revised question—that is, a method other than building a 30-foot pendulum. The task is to use mathematics to find the answer.

Key Questions

What do we know from the story?

What do we need to know to decide whether the prisoner can escape?

What information and materials do you need to answer the unit question?

POW 9: The Big Knight Switch

Intent

In this first POW of the unit, students examine a puzzle based on the movement of chess pieces.

Mathematics

As with many POWs, the most important content addressed in the activity is the problem-solving process itself. In this case, students will work to develop sequences of steps that move a chess piece from one part of the board to another. The problem raises the important mathematical questions of existence, uniqueness, and efficiency of solutions.

Progression

Students do initial explorations during class and then work on and complete a write-up of the activity on their own. Several students will present solutions in a week or so.

Approximate Time

10 minutes for introduction

1 to 3 hours for activity (at home)

15 minutes for presentations

Classroom Organization

Whole class, then individuals, followed by whole-class presentations

Materials

1-inch Graph Paper blackline master (optional, handouts and transparency)

Doing the Activity

Discuss in class how the knight piece moves on a chessboard to be sure students understand the kinds of moves allowed. The graph grids should be large enough to put knights or markers in the squares.

Discussing and Debriefing the Activity

Following the POW presentations, you might ask whether other students were able to move the knights in fewer moves, and if so, have them share their methods. (The switch is possible, and the least number of moves in which the switch can be made is 16.)

Ask students to discuss how they kept track of moves so that they did not double-count and why they feel they cannot do the task in fewer moves. They probably won't offer complete proofs here, but this is a good opportunity for learning to develop convincing arguments.

Key Questions

Did anyone switch the knights in fewer moves?

How did you keep track of the moves?

Why do you think your result is the best possible?

Supplemental Activities

More Knights Switching (extension) is a similar activity that is a bit more challenging and is a good follow-up to the discussion of *POW 9: The Big Knight Switch.*

A Knight Goes Traveling (extension) is similar to the other knight-switch problems but uses a square chessboard of some unknown size. This is another good follow-up investigation to *POW 9: The Big Knight Switch.*

Building a Pendulum

Intent

Students immerse themselves in the unit problem by building their own pendulums, trying to time one full swing, and then reporting their results to the class.

Mathematics

The **period** of a pendulum is the time it takes for the bob of a pendulum to swing from some point P along its entire path and back to P. The most convenient point P is the point of maximum displacement from the vertical. In this activity, students will devise their own techniques to try to measure a pendulum's period.

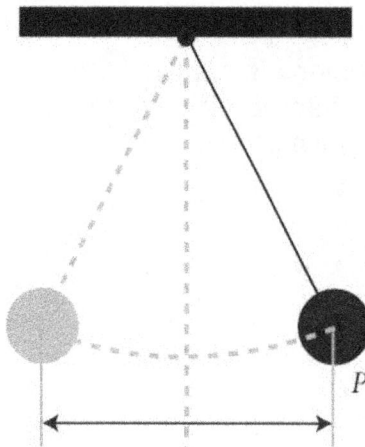

Progression

After a quick orientation, students will work on the activity at home and bring in their results to share with the class.

Approximate Time

5 minutes for introduction

25 minutes for activity (at home)

15 minutes for discussion

Classroom Organization

Individuals, followed by whole-class discussion

Doing the Activity

Have students brainstorm how to build a pendulum from materials found at home. It need not be elaborate, but as students will want to be somewhat accurate in measuring the period, talk about items they should consider using and not using.

© 2009 Interactive Mathematics Program

Discussing and Debriefing the Activity

Have students share their pendulums and what they have discovered in their groups, and then ask a few groups to report their findings to the class. Some students may have timed several swings and then divided by the number of swings; others may have measured several individual swings and averaged the results.

There is no single best way to measure the period of a pendulum; there are pros and cons to each method. For example, it is easier to make a single measurement of the time required for several swings than it is to do several measurements of one swing at a time. On the other hand, a student could argue that timing several swings in one measurement is not a valid method because the pendulum gradually slows down. If students do not have different approaches, there is no need to pursue this issue now; its importance will be examined in *What's My Stride?*

Key Questions

How did you measure the period of your pendulum?

What problems did you have in measuring the period?

Supplemental Activity

Getting in Synch (extension) presents the concept of *period* in a different context, through which students will explore some number theory. The activity can be assigned at any time, as the term period is introduced in the first activity. For whole-number periods, the questions are fairly straightforward, but for fractions, the issues get more interesting. Question 3 hints at an idea that Greek mathematicians called *incommensurate numbers*. This refers to a pair of numbers that we would today describe as having a ratio that is an irrational number. Don't expect students to get the correct answer to this question.

Initial Experiments

Intent

Students will follow their initial forays into the behavior of a pendulum by testing the effects of several variables on the period of a pendulum.

Mathematics

To answer the unit question, students will have to understand what variables affect the swing of a pendulum and how they affect it. In other words, they will be searching for functional relationships between **independent variables**, such as the pendulum's length, the angle of swing, and the weight of the bob, and the **dependent variable**, the period. To do this, they will perform a series of controlled experiments. In this early activity, they will compile a list of variables they think might affect the period and then each test one of these variables.

Progression

The activity begins with a discussion of variables that might affect a pendulum's period. Students then work in groups to test one of the variables, using methods of their own design. Finally, they report what they did and what they learned.

Approximate Time

35 minutes for activity

30 minutes for discussion

Classroom Organization

Groups

Materials

Unwaxed dental floss or fishing line

Rulers or metersticks

Heavy washers (all the same size and shape)

Stopwatches

Scissors

Doing the Activity

Remind students that to answer the initial question, *Does the story's hero really have time to carry out his escape plan?,* they must consider the narrower question that was posted yesterday, **How long would it take for Poe's pendulum to make 12 swings?**

Explain that, to answer this revised question, students will follow this general plan of investigation.

1. Find out which variable or variables affect the period of the pendulum.

2. Collect data on the periods of various pendulums, changing only the variable(s) decided on in step 1.

3. Look for patterns in the data, and use them to predict how long 12 swings of Poe's pendulum will take.

Post these steps under the revised question for students to refer to throughout the unit.

Ask groups to now consider this question: **What might affect the period of a pendulum?** Have groups make lists of potentially relevant variables, that is, of all the things they think might influence the period of a pendulum.

Then work together to make a class list of things that might influence the period of a pendulum. You can introduce the term *bob* for the weight at the end of the pendulum. Here are some variables students might propose.

- Weight of the bob

- Distance the bob is pulled back

- Composition of the bob (what it is made of)

- Shape of the bob

- Length of the string

Because this is simply a brainstorming process, use whatever ideas students present. For example, they might more naturally think about the distance the bob is pulled back than the angle of the pendulum with respect to the vertical. In this activity, it is fine if they use as their variable the horizontal distance the bob is pulled from its resting vertical position, or the curved distance through which the bob is pulled, instead of the angle of amplitude. Whatever measurement they use, though, be sure they define it precisely.

The general concept of **angle**, and the specific notion of *angle of amplitude* for a pendulum, will be introduced in *The Standard Pendulum* and related activities found in *Pendulum Variations.*

Students will conduct today's experiments in their groups. With this in mind, you may want to give some attention to this question: **How would you judge a group member's contribution?** To emphasize issues of group cooperation, you might ask the class to create a list of criteria on which this group work should be judged, such as the following.

- Does the group divide the work among the members?

- Does the group agree on a plan or structure for tackling the activity?

- Does the group take time to ensure that everyone understands the activity?

- Does the group use its time productively?

- Does the group support each member?

- Does the group record its results carefully?

You might post this list of "criteria for a successful group" for students to refer to. Once discussion of the activity is complete, you might return to these criteria as a follow-up. (Another opportunity for work on this area of learning is suggested for *Pendulum Conclusions*, in which students assign grades to themselves and to their group members based on their contributions to group work.)

In this activity, each group is to focus on a single variable. Assign a variable to each group, or let each group choose a variable from the list they generated. You may want to ensure that at least one group looks at each of the key variables of weight, length, and amplitude.

Provide an assortment of materials (whatever you have readily available) from which groups can choose. The activity is not intended to produce definitive results; instead, students should have the freedom to experiment and to make mistakes. In later activities—*The Standard Pendulum, Pendulum Variations,* and *An Important Function*—they will be performing other pendulum experiments with a better understanding of the process of experimentation.

As groups work, be sure they are recording their findings and planning how they will report their procedures as well as their data. Students will vary in their efficiency at completing their experiments, so allow enough time that all groups can compile at least some results. Encourage groups to explore other variables if they have time.

Discussing and Debriefing the Activity

As groups share their results, follow up on the earlier discussion of experimentation. Whenever possible, point out the difficulty in making a decision about "what matters" without keeping all variables constant except the one being studied. Some groups may have done fairly controlled experiments, changing just one variable. They may state, for example, that they changed the weight of the bob and got a different period. (This idea of a *controlled experiment* will be formally introduced in the next activity, *Close to the Law.*)

You can ask groups about their confidence in their results. **How certain are you that weight makes a difference? Are you 90% sure? Are you absolutely certain? Not at all certain?**

If the opportunity arises, ask whether groups ever measured the period of the same pendulum more than once and, if so, whether they got exactly same result. This discussion can be an introduction to the idea of **measurement variation**, a topic that will arise in a later activity.

Inform students that these initial experiments are not the last they will perform. However, before they conduct additional experiments, they need to study more about how to tell whether a variable "makes a difference" in the period of a pendulum's swing.

Key Questions

What variable did you measure?

What did you do to measure this variable's effect on the period of the pendulum?

How long do you think it will take a 30-foot pendulum to make 12 swings?

How certain are you that your variable affects the period?

Did you ever get different periods for the same pendulum?

Supplemental Activity

Height and Weight (reinforcement or extension) asks students to explore whether the weight of something is correlated with its height (in a context of the students' own choosing). This activity relates to today's discussion about what variables are related to the period of a pendulum's swing.

Close to the Law

Intent

Students identify variables that might affect the outcome of an experiment.

Mathematics

To determine whether there is a relationship between two aspects of some situation—such as the number of crimes on a city block and that block's distance from the nearest police station—one would conduct a controlled experiment. That is, one would hold all (or almost all) other aspects of the situation constant so as to isolate the effect of one of the two aspects in question on the other. One would then change one aspect and try to observe a change in the other, and then repeat this process. The resulting data would be analyzed to see whether they support the hypothesis that a relationship exists. This activity helps students refine their understanding of this process.

Progression

Students will work on the activity individually and then share their results in a class discussion.

Approximate Time

15 minutes for activity (at home or in class)

15 minutes for discussion

Classroom Organization

Individuals, followed by whole-class discussion

Doing the Activity

This activity will require little or no introduction.

Discussing and Debriefing the Activity

Have volunteers share their ideas about the activity. Some students might think that the data support the position that having more police stations would reduce crime, as, according to the chart, the number of crimes does rise as the distance from the police station increases. Others may be skeptical about this conclusion. Use their skepticism to explore why—given the fact that the number of crimes does increase with the distance—one might not conclude that building more police stations will decrease crime.

One relevant issue is whether the differences shown in the table are great enough to justify any conclusions. This question will be important later in the unit, so look for opportunities to discuss it. If it doesn't come up on its own, you might ask something like, **What if the numbers in the table were 1.3, 1.31, and 1.32? Would you reach the same conclusion?**

This discussion may blend into the discussion of Question 2. Other explanations for differences in crime rate might include lack of streetlights, presence of activities that attract crime, and a higher concentration of people.

Building on this idea, ask, **How could we determine whether the increase in crime is actually caused by increased distance from the police station?** Students might suggest, for example, comparing blocks that are essentially the same in other ways that might affect crime or the need to gather data over a longer period of time.

Introduce the term *controlled study* or *controlled experiment* to refer to what students are describing, and see if they can put this general idea into words. They might describe it as a study or an experiment in which everything is the same except for the variable being tested.

Key Question
How might you test whether distance from the police station is the reason for the increase in crimes per block?

Supplemental Activity
Octane Variation (reinforcement) is similar to this activity, asking students to assess the validity of potential conclusions based on a small data set and to propose improvements to the experiment.

If I Could Do It Over Again

Intent

Students reflect on their experiments of the last few days. In the process, they evaluate successes and rethink difficulties they encountered.

Mathematics

This activity gives students an opportunity to collect their thoughts about the difficulty of setting up and conducting controlled experiments.

Progression

Students will work on this reflective activity individually and then share their results in a brief class discussion.

Approximate Time

5 minutes for introduction

15 minutes for activity (at home or in class)

5 minutes for discussion

Classroom Organization

Individuals, followed by whole-class discussion

Doing the Activity

To introduce the activity, explain that in this unit, students will do many more pendulum experiments, enabling them to put their initial work—both what worked well and what worked not so well—to good use later.

Discussing and Debriefing the Activity

No two groups will have had identical successes and challenges with their first scientific experiments. Have students share with one another some of their "best practices."

Time Is Relative

Intent

Throughout the unit, students will be collecting and analyzing data. This is the first of three activities in which they conduct controlled experiments.

Mathematics

Students will see how much variation can occur in measuring a fixed phenomenon. In producing a set of data that approximates a normal distribution, they will encounter the important idea of **measurement variation**, the unavoidable imprecision when measuring quantities. They will also review the use of a frequency bar graph to display data.

Progression

Students will work in pairs to collect data. The class will pool the results to create a frequency bar graph.

Approximate Time

35 minutes

Classroom Organization

Pairs and small groups

Materials

Stopwatches (1 per pair of students)

Classroom clock or watches with second hands

Doing the Activity

The student book describes the specifics of this task: to time five seconds on a stopwatch based on the directions to start and stop from someone who is watching a clock or watch with a second hand or digital counter. The timer (holding the stopwatch) should not be able to see the clock or watch. Students take turns in their groups timing five seconds to the nearest tenth of a second and recording each result. Each timer should collect 10 pieces of data.

Once the data have been collected, work as a class to make a **frequency bar graph**, on chart paper, of all the data. You may need to review what a frequency bar graph is and how to make one. (Students were introduced to frequency bar graphs in *The Game of Pig*.) Essentially, they need to recall that this type of graph shows how often each particular result or range of results occurs.

Ask the class, **What might be a good way to group our data?** To decide how to group the data, students might suggest first finding the spread of the data, that is, the highest and lowest results in the class. You might remind them that each grouping should cover an equal range of possible results.

Discuss the pros and cons of various groupings. Students might recall from their experiences in *The Game of Pig* that seven or eight columns worked well to group a large set of data. The major point to bring out is that, if there are too many or too few subintervals, the graph won't look very interesting; they won't see any patterns in the graph.

Discussing and Debriefing the Activity

After the graph is complete, point out that students were trying to measure exactly the same thing but came up with a range of results. They may be surprised at how far from five seconds some of the results are. Tell students that this phenomenon is referred to as **measurement variation**.

Ask the class, **What does measurement variation have to do with the unit problem?** If they need a hint, point to step 1 of the plan of investigation from *Initial Experiments:*

1. Find out which variable or variables affect the period of the pendulum.

How does the idea of measurement variation affect your conclusions from your previous pendulum experiments? Bring out that the variations they saw among different pendulums might have been due only to measurement variation.

Ask them to briefly review their conclusions from those experiments. Tell them that they will conduct another set of experiments later on, when they have a fuller understanding of the issue of measurement variation.

Post the class graph for use in the introduction to normal distribution in the activity *What's Normal?*

Key Questions

What might be a good way to group our data?

What does measurement variation have to do with the unit problem?

How does the idea of measurement variation affect your conclusions from previous pendulum experiments?

What's Your Stride?

Intent

This is a second of three activities in which students conduct controlled experiments and display the results of the entire class.

Mathematics

This activity will give students additional experience with **measurement variation**. When measuring a given quantity, the measurement variation will be approximately normally distributed. That is, errors are as likely to underestimate as overestimate the "real" value of the quantity, and smaller errors are more likely than larger ones.

Progression

Students will gather data outside of class and then combine their results to create a frequency bar graph of the class data set.

Approximate Time

5 minutes for introduction

25 minutes for activity (at home)

25 minutes for discussion

Classroom Organization

Individually, then groups, followed by whole-class discussion

Materials

Metersticks, yardsticks, or measuring tapes

Doing the Activity

Introduce the activity, and talk about possible techniques for measuring stride length. Students may suggest various methods for measuring the length of a single stride, such as taking the length of several strides and dividing by the number of strides.

Discussing and Debriefing the Activity

When students return with their collected data, ask them to work in their groups on the following tasks.

- List all the methods for calculating stride length that the group members used.

- Discuss the strengths and weaknesses of each method.

Now work as a class to prepare a frequency bar graph, on chart paper, of the stride-length data collected by the entire class.

Ask, **How do you want to group the data?** Bring out again, if needed, that it is important that the intervals are of equal width; otherwise, the graph will give a misleading picture of the results. For example, if measurements are given to the nearest inch, the graph might include bars for stride lengths of 15 to 19 inches, 20 to 24, 25 to 29 inches, 30 to 34 inches, and so on.

Once the class has agreed on groupings, ask each group to total their results in each interval, and combine these totals to make the graph.

Using smaller or larger intervals may produce more interesting results. If time allows, make more than one frequency bar graph, using a different choice for the width of the intervals, and compare the results.

Post the graph for use in the introduction to normal distributions in *What's Normal?*

Key Question
How do you want to group the data?

Pulse Gathering

Intent

This is the third activity in which students collect measurement data. In *Pulse Analysis,* they will analyze their collected data.

Mathematics

Students will continue to encounter measurement variation as they gather pulse rate data. Variations in pulse measurements will be due to error and to actual changes in pulse rate, which is not constant over time even for a single individual. The challenge in this situation, and in the unit problem, is to distinguish between real differences and differences due to measurement variation.

Progression

This is a data collection activity. Students will measure and record their own pulses and then record the measurements for at least two other students.

Approximate Time

25 minutes

Classroom Organization

Individuals, then groups

Materials

Clock with second hand or stopwatch

Doing the Activity

Have students read the activity. Ask them how to find their pulse, and make sure all of them can find their pulses on their own wrists or necks. If students have not commented on the importance of using a finger rather than the thumb, caution them that thumbs have a pulse of their own.

Discussing and Debriefing the Activity

Tell students they will be working with the data from this activity in *Pulse Analysis*. For that activity, they will need data from at least two other students, so they should record the data from the other members of their groups.

Pulse Analysis

Intent

Students will analyze the data from *Pulse Gathering,* including, for the first time, calculating and considering the mean of a set of data and continuing to make and interpret frequency bar graphs.

Mathematics

This activity provides further insight into the concept of **measurement variation**. The **mean** of a set of data collected by measurement is the best estimate of the value being measured. In theory, if the measurements were repeated an infinite number of times, the mean would be the "true" value.

Progression

Students will begin this activity individually and then continue it in their groups and with the entire class.

Approximate Time

25 minutes for activity (at home or in class)

20 minutes for discussion

Classroom Organization

Individuals, then groups and whole class

Doing the Activity

Use this activity as an occasion to clarify, once again, the concept of measurement variation. You may want to have students think back to their work on previous activities, beginning with *Time Is Relative.*

Discussing and Debriefing the Activity

You might begin the discussion by focusing on why students didn't get the same pulse rate each time they measured. Bring out that the variation could have been due either to an actual change in pulse rate or to measurement variation.

Again, ask about the role of this phenomenon in the unit problem. **What role does measurement variation play in the unit problem?** Students should be realizing that measurement variation will be a crucial factor in their pendulum experiments. They will need to decide whether a change in the measurement of a pendulum's period is due to measurement variation or to the effect of a change in a variable.

Bring out that the data vary from person to person, and ask whether students could tell from a single measurement in their list which person the pulse rate came from. **Can you tell from one measurement whose pulse it was?** Even though different individuals probably have different means, there is most likely quite a bit of overlap in the results. For example, one person may have had measurements between 16 and 20, and another from 18 to 22, so a measurement of, say, 19 could have come from either person. Taking one measurement from each person wouldn't necessarily tell which person had the higher mean pulse rate.

Ask students, **Are there any observations you can make about how your group graph (from Question 3b) compares to your individual graphs (from Question 3a)?**

Compile a master chart of the class data. Use the groups' subtotals (from making the graphs for Question 3b) to compute the totals for the class. In a class of 30 students, there will be 300 individual data items.

You may want to discuss the spread of the data. If any results are far off, ask for possible explanations. This is a convenient time to introduce the term **outlier**. Although this term has a technical definition, it refers generally to an experimental result that seems far removed from the main body of the data.

Then ask each group to put the class data into a frequency bar graph, as done with the timing and stride data. Let groups make their own decisions about how to group the data.

Post one or more of these graphs to use in the introduction to the normal distribution in *What's Normal?*

Key Questions

What role does measurement variation play in the unit problem?

Can you tell from one measurement whose pulse it was?

Supplemental Activities

Stem-and-Leaf Plots and *Quartiles and Box Plots* are reference materials that summarize two methods of representing numeric data and introduce the concept of quartiles, an extension of the concept of median, which is used in creating box-and-whiskers plots.

Data Pictures (reinforcement) provides an opportunity for students to construct three different types of visual representations of a data set—a frequency bar graph, a stem-and-leaf plot, and a box plot—and examine the advantages and disadvantages of each. This activity can be used in conjunction with the reference pages *Stem-and-Leaf Plots* and *Quartiles and Box Plots*.

POW 10: Corey Camel

Intent

In this POW, students practice clear record keeping as an essential part of problem solving.

Mathematics

The most important mathematics addressed in this POW is the problem-solving process itself. First, students will have to try some initial guesses to get a sense of the structure of the problem situation. In a subsequent activity, they will consider a simpler version of the problem, thereby making explicit a powerful problem-solving strategy.

Progression

To help students simplify this large problem, a simpler version will be posed early in the next set of activities, in *A Mini-POW About Mini-Camel*.

Approximate Time

15 minutes for introduction

1 to 3 hours for activity (at home)

10 minutes to gauge progress

20 minutes for presentations

Classroom Organization

Individuals, followed by whole-class discussion and presentations

Doing the Activity

After students have read the activity aloud, model a specific case of Corey moving bananas. Mention that there are a variety of creative problem-solving strategies that might help students analyze Corey's situation.

Note that the POW refers to a simpler version of the situation, *A Mini-POW About Mini-Camel,* a class activity that will help students solve the larger Corey Camel dilemma.

Discussing and Debriefing the Activity

Have at least three students present their work on the POW. Then ask whether anyone was able to get more bananas to market than any of the presenters; if so, discuss their solutions. (If fractional bananas and miles are allowed, the best Corey can do is to get $533\frac{1}{3}$ bananas to market. If fractions are not permitted, Corey can get 533 bananas to market.) If no one has come up with the maximum possible number, mention that there is a better answer without revealing what that answer is or how to get it.

Ask the class, **How would you know when you have found the maximum possible number of bananas?**

Finally, discuss the role that the mini-POW played in their work on the POW itself. Here are some questions that might help to focus the discussion.

Was the mini-POW helpful? If so, how?

Why was the mini-POW easier to solve than the POW itself (if it was)?

What is special about the numbers in both versions of the camel problem?

Could you make up another version (Super Camel?) that you could solve automatically from what you already know?

Use students' comments to bring out the idea that insight into a hard problem might come by first looking at a simplified version of that problem.

Key Questions

Was anyone able to get more bananas to market?

How would you know when you have found the maximum possible number of bananas?

How did the mini-POW help with the POW?

Return to the Pit

Intent

This activity asks students to summarize where they are in the process of solving the unit problem and, specifically, to reflect on the role of measurement variation.

Mathematics

As this set of activities draws to a close, students reflect on the mathematical ideas raised so far: the idea of a controlled experiment, the variables that might affect the period of a pendulum, and the distribution of error when measuring.

Progression

Students will work on this reflective activity individually and then briefly share their results with the class.

Approximate Time

20 minutes for activity (at home or in class)

10 minutes for discussion

Classroom Organization

Individuals, followed by whole-class discussion

Doing the Activity

This is a chance for students to use their writing skills to clarify what they have been thinking about as they conducted their initial experiments and collected and analyzed data in the previous four activities.

Discussing and Debriefing the Activity

You might have a few volunteers share their views of what the problem is, how the work they have done so far might relate to its solution, and the role of measurement variation in their work. If students have questions about the unit or items they don't understand so far (Question 3), you might list the questions and ask whether anyone can help answer them.

Statistics and the Pendulum

Intent

By doing the activities in *Statistics and the Pendulum*, students begin to build an understanding of the tools they will need to test the effects of different variables on the period of a pendulum.

Mathematics

Of the many ways in which data may be distributed, one of the most useful—especially when dealing with data collected from measurements—is the **normal distribution** (sometimes called the bell curve). Given the role of measurement in addressing the unit problem, understanding the properties of this distribution will be important for students. The keys to measuring the variability in a normal distribution are two statistics: **mean** and **standard deviation**. When data are distributed normally, we know how much data resides within a certain number of standard deviations of the mean of the distribution.

In these activities, students learn how to calculate and interpret standard deviation and then use this statistic, along with the normal distribution, to reason about data.

Progression

Statistics and the Pendulum begins by defining the normal distribution. Students then reason informally about normally distributed data as they develop an understanding of standard deviation. The unit ends with students doing more formal reasoning about normally distributed data using standard deviation. The calculator will be introduced as a helpful tool for summarizing and displaying data. In addition, students will complete their work on the second of the unit's POWs and begin work on the third.

What's Normal?

A Mini-POW About Mini-Camel

Flip, Flip

What's Rare?

Penny Weight

Mean School Data

An (AB)Normal Rug

Data Spread

Kai and Mai Spread Data

Standard Deviation Basics

The Best Spread

Making Friends with Standard Deviation

Deviations

POW 11: Eight Bags of Gold

Penny Weight Revisited

Can Your Calculator Pass This Soft Drink Test?

What's Normal?

Intent
This activity, and the discussion that leads into it, introduces students to the normal distribution.

Mathematics
The **normal distribution** (also called the *Gaussian distribution*) is the technical name for what many call the *bell curve*. Of the many ways that data may be distributed, the normal distribution is of particular interest and is useful in many statistical situations. For example, many types of data related to people—such as the heights or shoe sizes of adult men or women—are approximately normally distributed. The normal distribution is a specific type of bell-shaped frequency pattern, with a precise, technical mathematical definition. Additionally, **measurement variation** is approximately normally distributed. It is for this last reason that the normal distribution is introduced in this unit.

Progression
After a teacher-led introduction to the normal distribution, students work individually to create graphs of surmised data from several situations, including labeled axes and their own choices for intervals, units of measurement, and frequency of data within each interval. They then share their results in groups.

Approximate Time
30 minutes for introduction

20 minutes for activity (at home or in class)

10 minutes for discussion

Classroom Organization
Whole-class introduction, then individuals, then groups, followed by whole-class discussion

Materials
Frequency bar graphs from *Time Is Relative, What's Your Stride?,* and *Pulse Analysis*

What's Normal? blackline masters (transparency)

Doing the Activity
Before assigning the activity, lead a discussion to introduce the normal distribution. To begin, draw students' attention to the frequency bar graphs made earlier of the following data sets.

- Timing of five seconds (from *Time Is Relative*)

- Stride length (from *What's Your Stride?*)

- Pulse rates (from *Pulse Analysis*)

Ask students, **What features do these graphs have in common?** They will probably focus on two key features.

- The graphs are highest "in the middle." (Students may or may not use the term **mean**.)

- The graphs gradually go down toward the ends.

Using a diagram like the one below, explain that curves with this general appearance are called *bell shaped* and that there is a very special bell-shaped curve called the **normal distribution**.

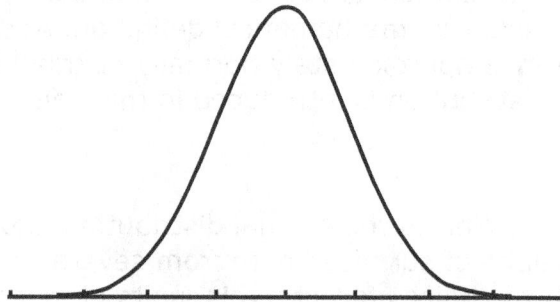

Bring out the connection between the area under such a curve and the probability of various results. For example, if the shaded area on the next diagram is, say, 20 percent of the total area under the curve, then 20 percent of all measurements are between points *a* and *b*.

Students may recognize that a similar idea applies to frequency bar graphs. Point out the similarity between this shaded area under the curve and the area of a bar in a frequency bar graph. It's as if the tops of all the bars in a frequency bar graph were connected to draw a smooth curve.

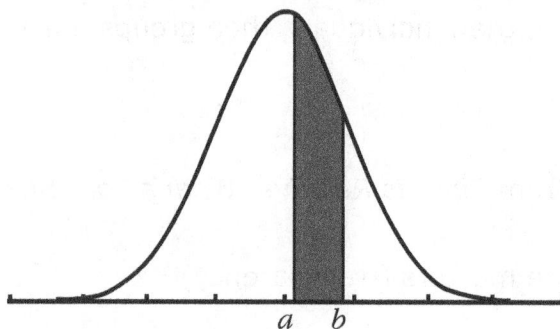

You can also show the next diagram, which depicts three different normal curves on the same set of axes, and ask students what they think the differences indicate.

The goal is for them to recognize that the amount of variation from one measurement to another is different in each graph. The exact shape of a normal curve depends on the scales being used and the specific situation.

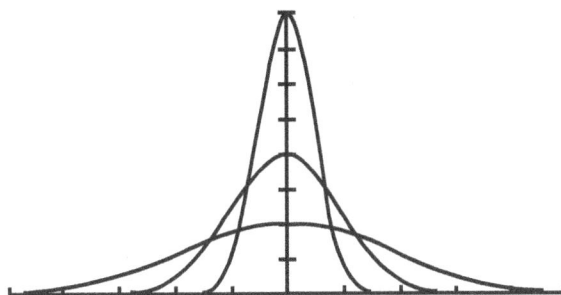

Tell students that you are giving them a simplified description of the normal distribution. They will not be able to determine for sure whether a data set is normally distributed. The precise definition involves a complex formula for the graph—one that most people encounter only if they study statistics in college.

You may want to clarify that in the term *normal distribution,* the word "normal" is being used in a special, technical sense. It does not mean "ordinary," although the normal distribution is one that occurs in many situations.

Ask, **What features do these normal curves have in common?** Students should see that, as with the frequency bar graphs under consideration, the normal curves are highest in the middle and decrease gradually toward both ends. Make sure they note one more specific phenomenon:

The normal curve is symmetric.

Introduce the term **line of symmetry** for the vertical line that divides a normal curve into two equal parts. Then ask, **What does the location of the line of symmetry represent?** Students should realize that values to the right of the line of symmetry "balance out" values to the left. Help them as needed to use this observation to reach an important conclusion:

The line of symmetry represents the mean of the data.

If students mention the median (in addition to or instead of the mean), explain that for symmetric data, the **mean** and **median** are the same, and perhaps ask why.

A more subtle observation on the shape of the graph concerns *concavity*. You can bring this out by asking, **What changes in the way the normal curve "curves"?** You might use the following diagram to illustrate the ideas. Introduce the terms *concave up* and *concave down* to describe the different portions of the curve.

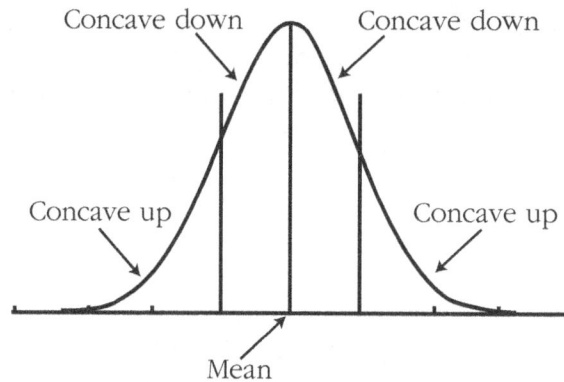

Note that the change of concavity provides an important visual image of standard deviation. For example, the point on the horizontal axis that corresponds to the first point of concavity to the right of the mean is one standard deviation above the mean. The significance of concavity in relation to standard deviation will be discussed in the activity *The Best Spread*.

Ask, **Do you think the frequency bar graphs of our experimental data resemble the normal distribution?** Students' response may depend on how much data they collected for each experiment and on how they grouped the results. Whatever their response, tell them that if they were to record more and more data, their graphs would probably begin to look more and more like the normal distribution. The normal curve is generally considered a reasonable expectation for results of measurement variation. Then tell them that, based on this general experimental phenomenon, this unit makes the following assumption:

Normality assumption: *If you make many measurements of the period of any given pendulum, the data will closely fit a normal distribution.*

Post this assumption in the room, as it will be referred to later in the unit.

Now ask, **How does the idea of normal distribution relate to the unit problem?** Students should recognize that, according to the normality assumption, the normal distribution describes the kind of measurement variation they should expect in pendulum experiments. Therefore, familiarity with the normal distribution is moving them along in the process of determining which variables are important.

Discussing and Debriefing the Activity

In their groups, have students compare the frequency bar graphs they sketched. Then discuss, as a class, which situations they think are normally distributed.

For Question 1, some students may have arranged the categories so that the tallest frequency bars are in the middle and then concluded that the distribution is approximately normal. If so, point out that a normal distribution requires that the data items be numeric in nature. Tell the class that nonnumeric data, like shoe type, is sometimes called *categorical data*.

Of Questions 2 through 4, only the situation in Question 2 might be approximately normally distributed (and even that might not be), although students may not have the facts on which to make this judgment.

For Question 3, bring out the fact that far more people have incomes below the mean than above it (due to the effect on the mean of a small number of people with very high incomes). In particular, this means that the distribution of incomes is not symmetric around the mean. As needed, review that symmetry is one of the key characteristics of the normal distribution. However, income distribution does resemble the normal distribution in at least one respect: It trails off toward the extremes (at least at the upper end). You can review here that in the normal distribution, values farther from the mean are less likely (that is, occur less often) than values closer to the mean.

For Question 4, help students to see that, assuming either constant or increasing birthrates, the population of different age groups decreases gradually toward the higher age groups. For example, there are generally more people between the ages of 0 and 10 than between 10 and 20, more between 10 and 20 than between 20 and 30, and so on. (States like Hawaii and Florida, which attract many retirees, might be exceptions to this pattern.)

You might mention that many properties of people and objects are distributed normally or close to normally, but many are not. It isn't necessarily easy to decide, in theory, which are which.

You may want to summarize several key aspects of the normal distribution that were brought out in this activity.

- Normally distributed data must be numeric.

- Normally distributed data are symmetric about the mean.

- For normally distributed data, results farther from the mean are less likely than results closer to the mean.

Finally, review the assumption that is being made in this unit: that measurements of a given pendulum's period are normally distributed.

Key Questions

What features do these graphs have in common?

What does the location of the line of symmetry represent?

What changes in the way the normal curve "curves"?

Do you think the frequency bar graphs of our experimental data resemble the normal distribution?

How does the idea of normal distribution relate to the unit problem?

Which situations do you think are normally distributed? Why?

A Mini-POW About Mini-Camel

Intent

This activity encourages the application of problem-solving strategies that will assist students in their work on *POW 10: Corey Camel.*

Mathematics

The problem-solving strategy "solve a simpler problem" is the focus here. Students may also employ the "make a drawing" strategy. Students can transfer their understanding of this simplified problem to the larger situation presented in *POW 10: Corey Camel.*

Progression

Students will work on this task in small groups and then present their results to the class.

Approximate Time

35 minutes

Classroom Organization

Groups, followed by whole-class discussion

Materials

45 beans or counters per group (to represent the 45 bananas)

Doing the Activity

Tell the class that they will apply what they learn from today's mini-POW to the POW itself. Remind them to keep notes on how they solve this problem. Tell students who have already solved the POW that they can give hints to or ask questions of their group members about how to solve this task. Emphasize that they should not tell their group mates the solution, but just offer subtle suggestions or ask good questions regarding types of things to think about.

Discussing and Debriefing the Activity

Ask volunteers to present their solutions to the mini-POW, asking questions to help make their explanations clear. You might delay a discussion of how the mini-POW specifically relates to the POW. However, you can mention that trying to improve on their solution will help students in their work on the larger problem.

The best possible result for Mini-Camel is to get eight bananas to market. If no one has found this result, tell students that there is a better answer, but not what it is or how to get it.

Key Questions

What helps to make an explanation clear?

How might a diagram along with your written explanation be more convincing?

Flip, Flip

Intent

In this activity, students conduct experiments in a situation in which the results are normally distributed.

Mathematics

Students began to collect and analyze experimental data in *The Game of Pig*. Now they are reminded that there is a difference between experimental and theoretical results, but that these differences are reduced as the number of trials increases. Students will encounter the coin-flip distribution in later years of IMP. At this time, it is enough that the discussion supports their understanding that the distribution is approximately normal.

Progression

Students will work individually to collect data and then pool their results with their group mates and, ultimately, with the entire class.

Approximate Time

5 minutes for introduction

25 minutes for activity (at home or in class)

20 minutes for discussion

Classroom Organization

Individuals, then groups, followed by whole-class discussion

Materials

10 coins per student

Doing the Activity

Demonstrate that a single experiment consists of flipping a set of 10 coins and counting the number of heads. Emphasize again, as students learned in *The Game of Pig,* the importance of gathering genuine data on this and similar assignments.

Discussing and Debriefing the Activity

Ask students for verbal descriptions of their frequency bar graphs. They will probably offer a simple description, such as that most results are "in the middle" with fewer results at the extremes. (Note that because the total number of trials is relatively low, coin-flip results will not always match the theoretical probabilities.)

Ask each group to total the number of times each possible result occurred among its members (how many 0s, how many 1s, and so on). Have group representatives read the totals out loud while you record a master list on the board. Then create a frequency bar graph of the class data.

It will be interesting to see whether any 0s (no heads) or 10s (all heads) occurred, as well as what students' expectations about the results were. (In a class of 30 students, the probability is about 36% that at least one person will get a 0, about 36% that at least one person will get a 10, and about 59% that at least one of these two extreme cases will occur.)

For Question 4, students are likely to think that the graph will look normal. In fact, the theoretical distribution of the coin flips is quite close to normal. The graph below, created using Fathom Dynamic Data™ software, shows that with 10,000 trials the results approximate a normal distribution.

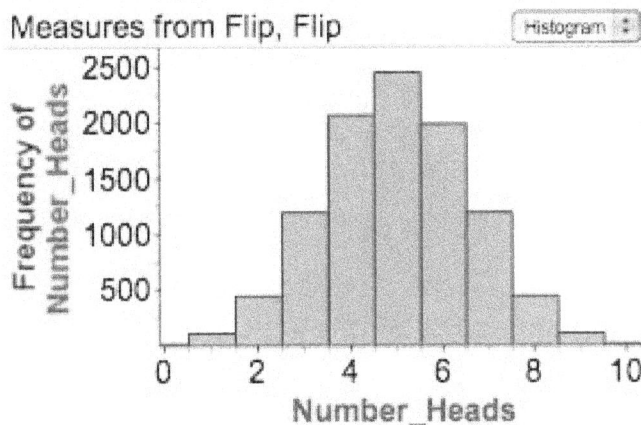

Question 5 is important as it exposes students to the idea of what conclusions might be drawn from unusual results. Ask students to discuss how much **deviation** you would have to see in order to decide that the coin was really unbalanced.

If anyone got a result in the coin-flipping experiment of 0 or 10 (or even 1 or 9), you might ask if that person suspects that the coins he or she used might have been unbalanced.

Through these experiments and discussion, students should be realizing that they can generally expect some deviation from the average when they conduct an experiment. They should also be gradually developing intuition about what level of deviation should be regarded as significant.

Key Questions

What does your frequency bar graph look like?

How does the class graph compare to what you predicted in Question 4?

What's Rare?

Intent

Students examine the results from three previous experiments to help them develop intuition about what is "ordinary" and what is "rare," which will help them to understand the concept of standard deviation later in the unit.

Mathematics

Standard deviation is one of many measures of the variability in a set of data. With normally distributed data, the standard deviation can be used to determine whether a particular result is an ordinary or a rare occurrence. For example, when flipping 10 coins, getting 4, 5, or 6 heads is an ordinary result, as each is within one standard deviation of 5, the mean value. However, getting 9 or 10 heads, or 0 or 1 heads (each more than two standard deviations from the mean), would be rare. In this activity, the notions of "ordinary" and "rare" are introduced in an intuitive way that will lay the foundation for understanding the meaning of standard deviation later in the unit.

Progression

Students will work on the four parts of this activity in their groups. The first three parts refer to results of earlier experiments, and the fourth asks students to step back and review their work across experiments.

Approximate Time

25 minutes

Classroom Organization

Groups, followed by whole-class discussion

Materials

Frequency bar graphs from *Time Is Relative, What's Your Stride?* and *Pulse Analysis*

Doing the Activity

Tell the class that in addition to providing numeric answers to the questions in this activity, they should keep track of the reasoning behind their answers.

The two categories of ordinary and rare need not cover all possibilities; students may decide that some results fall between the two.

Discussing and Debriefing the Activity

Begin the discussion by asking, **How did you decide what to call ordinary and what to call rare? How did you decide on the percentages?** The distinction between ordinary and rare is an arbitrary one, so have groups share how they decided to define these terms.

Now ask, **Were your percentages similar for all three situations?** Discuss whether the percentages defining ordinary and rare for one set of data hold up for another set of data—that is, whether there is some consistency in what students hold these terms to mean. If so, you will be able to build on this result when students look at standard deviation.

Finally, you may also want to discuss the question of whether it was the timing instruments or the person doing the timing that was responsible for the variation in the results for the experiment in *Time Is Relative*.

Ask students how their work of the last few days relates to the unit problem. They should mention that they have been gaining insight into the normal distribution—in particular, what level of variation can be considered ordinary and what level can be considered rare in normally distributed data.

Assure students that they will be applying everything they are learning to the pendulum question.

Key Questions

How did you decide what to call ordinary and what to call rare?

How did you decide on the percentages?

Were your percentages similar for all three situations?

How are these ideas connected to the unit problem?

Penny Weight

Intent

In this activity, students use their developing intuition about measurement variation to make a decision.

Mathematics

Given the weights of a collection of real coins, is another coin of a particular weight real or counterfeit? In other words, is the weight of the suspect coin what you would expect, given the variability in the weights of real coins? These are essential statistical questions. At this point in the unit, students will use frequency bar graphs and calculations of mean, median, and mode to address these questions, creating the need to develop a new tool later in the unit: standard deviation. Students will revisit Question 2 of this activity in *Penny Weight Revisited,* with the tool of standard deviation in mind.

Progression

Students will work on this activity individually and then share their results in a class discussion.

Approximate Time

5 minutes for introduction

15 minutes for activity (at home or in class)

20 minutes for discussion

Classroom Organization

Individuals, followed by whole-class discussion

Doing the Activity

Read through the activity as a class. Suggest that students may want to use statistics like the mean, median, and mode, as well as frequency bar graphs, to analyze the situation and to support their opinions.

Discussing and Debriefing the Activity

To begin the discussion, ask students if they think that the weights of pennies are normally distributed. (This specific set of data is not, but it is likely that the variation among pennies, though slight, creates an approximately normal distribution.)

Then ask, **Do you think Uncle Jack's coin was counterfeit? What is your reasoning?** Of those who do not think the coin was counterfeit, ask what weight it would have to be to convince them that it was.

Explain that although statistics will not tell whether the coin actually was counterfeit, statistics can reveal something about the probabilities involved. That is, it will tell them how rare such a weight is for a legitimate coin.

Bring out that there is no mathematical way to take Uncle Jack's personal reliability into account. Even without this sort of factor, there is often a subjective component to interpreting statistics.

Key Questions

Do you think the weights of pennies are normally distributed?

Do you think Uncle Jack's coin was counterfeit? What is your reasoning?

Mean School Data

Intent

In this activity, students bring their developing understanding of data distributions to bear on fictitious data related to the unit problem.

Mathematics

Two sets of data can have the same mean (or median or mode, for that matter), yet look very different otherwise. In this activity, students are given fictitious pendulum data from two high schools. The two sets of data have the same mean, but one set is more dispersed than the other. Two measurements of a data set—**center** and **spread**—will help students decide whether a particular observation is ordinary or rare.

Progression

Students will work on the two questions in their groups and then share their results with the class.

Approximate Time

30 minutes

Classroom Organization

Groups, followed by whole-class discussion

Doing the Activity

Spend a few minutes brainstorming some ideas that groups should consider as they analyze the two sets of data: mean, median, and mode, as well as what the data look like when they are graphed.

Discussing and Debriefing the Activity

Begin the discussion by asking, **What are the two means? How well do they represent each data set?** Students should recognize that there are some limitations to the use of the mean. Although the two sets of data have the same mean (1.22 seconds), they also have differences that the mean does not show.

Ask, **How do the two sets of data differ?** Try to elicit terminology such as "more spread out" or "more dispersed" to describe the King data in comparison to the Kennedy data.

As part of the discussion of Question 2, you might ask, **Which periods do you consider ordinary? Which periods do you consider rare?**

Key Questions

What are the two means? How well do they represent each data set?

How do the two sets of data differ?

Which periods do you consider ordinary? Which periods do you consider rare?

An (AB)Normal Rug

Intent

This activity will help students to think of the normal curve in terms of the areas it defines and the probabilities associated with those areas.

Mathematics

In this activity, students will consider the area under the normal curve and begin to develop an intuitive sense of where, for example, two-thirds or three-fourths of the data reside. This lays the groundwork for understanding another important property of standard deviation. In a theoretical normal distribution, about two-thirds of the area falls within one standard deviation on either side of the mean, and about 95 percent of the area falls between two standard deviations on either side of the mean.

Progression

Students will work on this activity individually and then share their results in a class discussion.

Approximate Time

10 minutes for activity (at home or in class)

10 minutes for discussion

Classroom Organization

Individuals, followed by whole-class discussion

Materials

An (AB)Normal Rug blackline masters (optional)

Doing the Activity

You might suggest that students trace the five diagrams in this activity onto graph paper or, alternatively, you can distribute copies of them.

Discussing and Debriefing the Activity

Use Question 1 to verify that students understand the idea of area under a curve.

Then turn to Questions 2 and 3. **How did you estimate where to draw the vertical lines?** You might let students use a transparency of the diagrams to share their estimates of where to draw the lines for each graph. Ask them to explain how they made their estimates.

Key Question

How did you estimate where to draw the vertical lines?

Data Spread

Intent

In this activity and the next, *Kai and Mai Spread Data,* students explore several ways to measure, in a quantitative way, the spread of a data set. These initial explorations prepare students for the introduction of standard deviation as a measure of spread.

Mathematics

Several statistics measure the spread of a data set. In this activity, students will be finding the **range** of the data: the numeric difference between the highest and lowest values in the set. As it uses two data points, this statistic is easily calculated. And two very differently distributed data sets can have the same range.

Progression

Students will work on *Data Spread* in class, and then continue with *Kai and Mai Spread Data* individually. Discussion in groups and as a whole class will follow.

Approximate Time

40 minutes

Classroom Organization

Groups

Doing the Activity

Introduce the activity by having the class examine and comment on the four sets of data. Bring out that the sets all have the same mean (the medians are almost the same as well) and differ mainly in how "spread out" they are.

You might also want students to consider how the data sets would look if they were displayed in frequency bar graphs. If you think it will help them, suggest they also calculate the mean, median, and mode for each set.

Discussing and Debriefing the Activity

The discussion should raise the observation that the range doesn't take into consideration the spread of data "inside" the least and greatest values.

Key Question

Which data set seems to be the most spread out from the mean?

Kai and Mai Spread Data

Intent
Students continue exploring methods to measure the spread of a data set. In this activity, they calculate the sum of the absolute deviations from the mean.

Mathematics
As a measure of spread, the sum of absolute deviations from the mean has an advantage over the range in that it uses every data value, or datum, in a set. The sum of the absolute deviations divided by the number of data values is called the **mean absolute deviation**. This statistic gives the average deviation from the mean of each data value.

Progression
Students will work individually on the activity and then share ideas and discuss the concepts of *absolute value* and *mean absolute deviation* as a class.

Approximate Time
25 minutes for activity (at home or in class)

25 minutes for discussion

Classroom Organization
Individuals or groups, followed by whole-class discussion

Doing the Activity
Assign this activity immediately after the discussion of *Data Spread*.

Discussing and Debriefing the Activity
You might have students convene in their groups to compare their personal schemes for measuring spread (Question 4) as well as the numeric results from the various methods. Then ask each group to report on the method they like best.

If students had trouble finding the appropriate numbers for each data set for Tai's, Kai's, and Mai's methods, review the mechanics of each method.

Then have the class discuss this question: **Which data set seems to be the most spread out from the mean?** The decision does not have to be made based on a statistical test, and the class need not reach agreement. Focus the discussion on *why* the class thinks a particular set is the most spread out.

You might ask why students think Mai chose to remove the highest and lowest values. One possible reason is that Mai wants to eliminate the possibility of one or two extreme values skewing the results. If there are any competitive divers in class, you might ask about the scoring for diving competitions (the highest and lowest scores are not considered).

Discussion of spread from the mean provides a nice opportunity to talk about **absolute value.** For example, ask someone to describe how to apply Kai's method to data set C: 12, 13, 13, 27, 27, 28. Use values above and below the mean (which is 20) to illustrate that in some cases you subtract the mean from the data item and in other cases you subtract the data item from the mean.

Then ask, **How could you find the distance from the mean the same way in all cases?** If needed, suggest that students consider the idea of absolute value. Point out that $|x - 20|$ (or $|20 - x|$) always gives a positive answer and measures how far a number x is from 20, whether x is greater than 20 or less than 20.

Introduce the symbol $\bar{x}$ (read "x bar") for the mean, and use notation such as x_i for a single piece of data. Then ask how one could write Kai's method using this notation and the summation symbol.

Students often enjoy seeing all this notation put together, using the summation notation developed in the *Patterns* unit, as

$$\sum_{i=1}^{n} \left| x_i - \bar{x} \right|$$

Later you can compare this notation with the symbolic formula for standard deviation.

Point out that, if Kai's number is divided by n, it gives the average deviation from the mean; it tells, *on the average,* how far the numbers in a set are from their mean. Introduce the term **mean absolute deviation** for this average, and have students express it using summation notation as

$$\frac{\sum_{i=1}^{n} \left| x_i - \bar{x} \right|}{n}$$

Try to get students to articulate what this expression measures. It is a simplified version of what **standard deviation** will tell them, which they will learn about next.

Students will need their work on *Kai and Mai Spread Data* for use during the next activity, *The Best Spread.*

Key Questions

Which data set seems to be the most spread out from the mean?

How could you find the distance from the mean the same way in all cases?

Standard Deviation Basics

Intent

These reference pages summarize the basic ideas about standard deviation.

Mathematics

The text defines **standard deviation** as a measure of how spread out a data set is and then discusses the following aspects of standard deviation:

- Calculating standard deviation

- Interpreting the standard deviation of normally distributed data

- Relating standard deviation to the concavity of the normal curve

Here are two reasons statisticians don't generally use mean absolute deviation—that is, just finding the average deviation from the mean:

- Mean absolute deviation involves the use of absolute value, which is difficult to work with in calculus. Squaring differences from the mean and then taking a square root at the end is computationally easier than using absolute value.

Mean absolute deviation is more suitable if the median is used as the "central value" instead of the mean.

Progression

The teacher leads a class discussion of the information in these reference pages.

Approximate Time

15 minutes

Classroom Organization

Whole class

Materials

Standard Deviation Basics blackline master (transparency, optional)

Discussing the Reference Pages

You may want to have students read over this material as part of their homework.

Define **standard deviation** using the example from the reference pages: the numbers 5, 8, 10, 14, and 18. (You may want to copy and hand out these pages so students can write on them.)

Calculation of Standard Deviation

Have students follow along with the steps and example in the student book as you demonstrate how to calculate the standard deviation.

1. Find the mean.

2. Find the difference between each data item and the mean.

3. Square each of the differences.

4. Find the average (mean) of these squared differences.

5. Take the square root of this average.

The computation of the mean is shown below the table to the left. Students may choose to ignore the sign of each difference in step 2, in effect using the absolute value of the difference rather than the difference itself. Since the differences are squared in step 3, their signs do not affect the final result. You may want to bring this out to emphasize the similarity between *standard deviation* and *mean absolute deviation*.

Below the table to the right, step 4 of the computation of standard deviation is broken down into substeps: (1) adding the squares of the differences and (2) dividing by the number of data items.

x	$x - \overline{x}$	$\left(x_i - \overline{x}\right)^2$
5	−6	36
8	−3	9
10	1	1
14	3	9
18	7	49

sum of data items = 55

number of data items = 5

$\overline{x} = 55 \div 5 = 11$

sum of data squared differences = 104

mean of the squared differences = 20.8

σ (standard deviation) = $\sqrt{20.8} \approx 4.6$

If students are comfortable with the summation notation presented in *Kai and Mai Spread Data* in the formula for mean absolute deviation, show them that the definition of standard deviation involves just two changes to that expression:

- Replacing $\left|x_i - \overline{x}\right|$ with $\left(x_i - \overline{x}\right)^2$

- Taking the square root of the final expression

Thus, the formula for standard deviation is

$$\sqrt{\frac{\sum_{i=1}^{n}\left(x_i - \overline{x}\right)^2}{n}}$$

The symbol usually used for standard deviation is the lowercase form of the Greek letter *sigma,* written σ. Mention this symbol, as students will be looking for it on their calculators. Remind the class that the uppercase Greek sigma, $\sum$, is the summation symbol. (The issue of the distinction between σ and s—the sample standard deviation—is discussed later.)

Standard Deviation and the Normal Distribution

Explain that the following facts hold true whenever a set of data is normally distributed:

- Approximately 68% of all results will be within one standard deviation of the mean.

- Approximately 95% of all results will be within two standard deviations of the mean.

Illustrate these facts using the areas shown in the next diagram, which also appears in the student book.

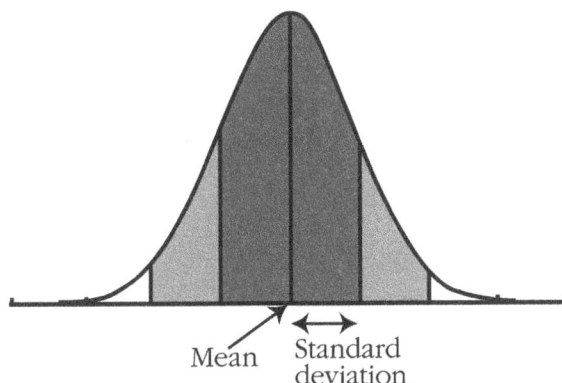

Mean   Standard deviation

The darkly shaded area stretches from one standard deviation below the mean to one standard deviation above the mean and is approximately 68% of the total area under the curve. The two lightly shaded areas represent data between one and two standard deviations from the mean.

The total shaded area stretches from two standard deviations below the mean to two standard deviations above the mean and comprises approximately 95% of the total area under the curve.

Explain that the standard deviation provides a good rule of thumb for deciding whether something is "rare."

The Best Spread

Intent

Students revisit their work on *Data Spread,* using the new tool of standard deviation. Their work will offer a sense of how comfortable they are with both standard deviation and the general idea of data spread.

Mathematics

Standard deviation is the most widely used statistic for measuring the spread of a data set. In this activity, students will revisit data for which they have already tried several other measures of spread, using standard deviation. Question 4 highlights the idea that different measures of spread can give different pictures of a distribution. For example, the data set 4, 4, 7, 9, 9 has a range of 5 and a standard deviation of about 2.24, while the set 3, 7, 7, 7, 10 has a greater range of 7 but a smaller standard deviation of about 2.23.

Progression

Students work together to solidify their ability to calculate standard deviation, work individually on the other questions, and then share their results in a class discussion.

Approximate Time

10 minutes for introduction

15 minutes for activity (at home or in class)

15 minutes for discussion

Classroom Organization

Whole class, then individuals, followed by whole-class discussion

Materials

Students' work on *Data Spread*

Doing the Activity

Have students work together to compute the standard deviation for one set of data before they work on their own.

Discussing and Debriefing the Activity

You might have students check their calculations for Question 2 in their groups. If there is disagreement, have volunteers present the calculations for the four sets of data. Then decide whether discussion of Question 3 seems needed.

A discussion of Question 4 is also optional. If you want to give students a hint for how to approach the problem, suggest they start with a particular data set and look for ways to change it

- first, so that the range stays the same but the standard deviation gets smaller, and

- second, so that the standard deviation stays the same (or at least stays smaller than that for the original set) but the range increases.

Make sure everyone understands how to calculate standard deviation.

Wrap up this activity by returning to the reference material *Standard Deviation Basics* and discussing the geometric interpretation of standard deviation. Students first encountered the ideas presented here in their work on *An (AB)Normal Rug,* so you might ask them to review their work on that activity before discussing the material in the reference pages.

Supplemental Activity

Data for Kai and Mai (reinforcement) poses questions similar to Question 4 of *The Best Spread*.

Making Friends with Standard Deviation

Intent
Students will continue to build their understanding of mean and standard deviation.

Mathematics
When a nonzero constant is added to each value in a data set, the **mean**, a measure of the center of the distribution, will change by that amount. However, the **standard deviation**, a measure of the spread of the data, will not change. When each value in a data set is transformed by multiplication by a particular constant, both the mean and the standard deviation will change. In this activity, students explore these transformations and draw conclusions about their effects on these two descriptive statistics. They will also continue to create data sets that have given values for these statistics.

Progression
Students work on the activity in groups and discuss their results as a class.

Approximate Time
35 minutes

Classroom Organization
Groups, followed by whole-class discussion

Doing the Activity
This activity requires little or no introduction.

Question 3 is intended primarily as further work for groups that finish early. You might warn students that they will be able to create data sets with those exact means but will only be able to approximate the standard deviations.

Discussing and Debriefing the Activity
Focus the discussion on parts c and d of Questions 1 and 2.

Students' explanations of the patterns they observe in Question 1d may take several forms. For example, they may picture the data points on the number line, so that adding the same thing to each data point just moves the points along and hence also moves the mean. Or they may see the change in the mean algebraically (although it's unlikely they will have a full algebraic explanation involving the distributive law).

Students may attribute the lack of change in the standard deviation to the fact that the spread doesn't change when the set of data points is moved along. Or they may recognize that when all the data are changed the same way, the mean also changes, so the spread from the mean remains the same.

The explanations for Question 2d will be similar.

Fathom Dynamic Data™ software can be used to provide a visual demonstration of the effects on a simple data set of adding or multiplying by a constant.

Key Question

What conclusions did you reach about how standard deviation is affected by changes in the data?

Supplemental Activity

Making Better Friends (extension) provides additional opportunities for students to explore standard deviation, continuing the challenges posed in this activity.

Deviations

Intent

In this activity, students continue to develop their understanding of standard deviation.

Mathematics

Students will apply their understanding of **standard deviation** as a measure of spread and the ideas developed in *Making Friends with Standard Deviation* related to the effects of linear transformations of data (adding a constant or multiplying by a constant) on the mean and standard deviation of a data set. In the activity, they will create new data sets with specific statistics as compared to a given data set.

Progression

Students will work on the activity individually and share their results in a class discussion.

Approximate Time

25 minutes for activity (at home or in class)

15 minutes for discussion

Classroom Organization

Individuals, followed by whole-class discussion

Doing the Activity

Point out that the instructions for Questions 2, 3, and 4 suggest that students need not actually calculate the mean and standard deviation for their new data sets. Instead, they can explain why they believe they have maintained the mean while changing the standard deviation or kept the standard deviation the same while changing the mean.

Discussing and Debriefing the Activity

You might first have students share their data sets in their groups and then ask for volunteers to describe their methods for creating the required data sets without calculating the means or standard deviations.

The mean of the data in Question 1 is exactly 20; the standard deviation is about 3.9. There are various ways to do Questions 2 through 4 without calculations. One approach to Questions 2 and 3 is to move the highest and lowest values either both away from the mean or both toward the mean by the same amount. For Question 4, one method is to add a specific amount to every data item.

Key Question

How does this activity relate to yesterday's activity, *Making Friends with Standard Deviation?*

Supplemental Activity

Mean Standard Dice (reinforcement or extension) offers practice with standard deviation. It also demonstrates that doing the same experiment more times doesn't change either the mean or the standard deviation. Students may be surprised about the result for standard deviation in Question 3.

POW 11: Eight Bags of Gold

Intent

This POW gives students another opportunity to solve and formally present solutions to an extended problem.

Mathematics

The context of this activity relates to some of the unit's tasks involving counterfeit or unbalanced coins. The most important content addressed by the POW is the problem-solving process itself. Students will try a variety of strategies for finding the one light bag of gold among eight bags using just a balance scale. They will search for a method that requires fewer than three weighings, and they will try to justify the claim that their method requires the fewest weighings.

Progression

Students are introduced to the POW. They present solutions in a week or so, with an opportunity for revision prior to that.

Approximate Time

15 minutes for introduction

1 to 3 hours for activity (at home)

20 minutes for presentations

Classroom Organization

Individuals, then groups, followed by whole-class presentations

Doing the Activity

Ensure that everyone knows what a pan balance is and how it is used to compare weights.

Students are scheduled to share their POW write-ups and write reviews of one another's work in the activity *POW Revision*. They will then revise their POWs (if they want to), and presentations will follow that.

Discussing and Debriefing the Activity

Have at least three students present their work on the POW to the class. Here is one strategy someone might present.

If the first weighing compares three of the bags to three of the others, one of two outcomes is possible:

- The two sets of three bags balance each other. This means the light bag is one of the two not yet examined, and a single additional weighing, comparing these two, is all that is needed to find it.

- One of the groups of three is lighter than the other. This means that the light bag is in this group. Only one additional weighing, comparing any two of these three bags, is needed. If they balance, the third one is the lighter one. If they don't balance, the lighter one will be evident.

Key Questions

Are you sure your scheme will always work?

If you got stuck working on this problem, how did you get unstuck?

Penny Weight Revisited

Intent

Students return to their work in *Penny Weight,* bringing to bear their new tool for judging ordinary and rare data.

Mathematics

In *Penny Weight,* students addressed these important mathematical questions: *Given the weights of a collection of real coins, is another coin of a particular weight real or counterfeit? Is the weight of the suspect coin what you would expect, given the variability in the weights of real coins?* At that time, students had few tools at their disposal to address these questions. Now they have standard deviation.

Progression

Students will work on the activity individually and then use their group members' work to evaluate their calculations and draw conclusions. After this activity, students will rely on their calculators, or software such as Fathom Dynamic Data software, to calculate statistics such as mean and standard deviation.

Approximate Time

30 minutes for activity (at home or in class)

45 minutes for discussion

Classroom Organization

Individuals, then groups, followed by whole-class discussion

Materials

Penny Weight Revisited blackline master (transparency, optional)

Doing the Activity

Advise students to keep track of all their computations so they can compare their results with others in their groups and locate any computational errors fairly easily.

Discussing and Debriefing the Activity

Have students work in their groups to achieve consensus on the answer to Question 1 and to begin sharing ideas about Question 2.

Once students agree on the value of the mean (2500 mg) and the standard deviation (approximately 9.88 mg), turn to answers and explanations for Question 2. The goal is for students to recognize that the weight of the uncle's penny is more than two standard deviations from the mean and that the chance of getting a result this far (or farther) from the mean is less than 5%. That is, if a coin were selected at random from a set of fair coins, the probability that it would be this far or farther from the mean is less than .05.

Note that the percentages for the normal distribution are essentially probability values. For instance, one could restate the condition given in *Standard Deviation Basics* that "Approximately 68% of all results are within one standard deviation of the mean" as "If a data value is chosen at random, the probability that it is within one standard deviation of the mean is approximately .68." Using the language of probability may help students connect the ideas in this unit to their work in *The Game of Pig*.

Ask the class what assumption they are making in using this value of 5%. They should see that they are assuming that the distribution of weights among pennies is approximately normal. You can assure them that this is a reasonable assumption. Remind them that the percentages in *Standard Deviation Basics* refer only to normal distributions.

Ask students, **Does 9.9 mg seem reasonable for the standard deviation?**

As needed, use the question to review the idea that standard deviation measures, more or less, a kind of average distance from the mean (although it is not literally an average). Students should agree that, because some pennies are less than 10 mg from the mean and others are more than 10 mg from the mean, a standard deviation of 9.9 mg seems about right. You might ask them to actually calculate the average distance from the mean—that is, the mean absolute deviation—which is a variation on Kai's method from *Kai and Mai Spread Data*. (For the pennyweight data, the value of mean absolute deviation is 8.2 mg.)

Ask students to sketch a normal distribution with a mean of 2500 and a standard deviation of about 9.9 and to shade the 5% area that they are discussing—that is, the area that represents results that are more than two standard deviations from the mean. The diagram should look something like this.

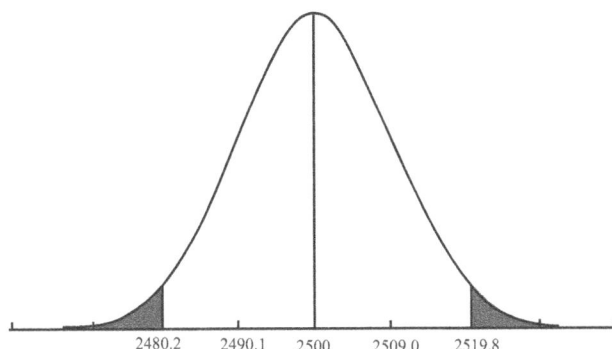

| | 2480.2 | 2490.1 | 2500 | 2509.0 | 2519.8 | |

Have students point out the regions on their sketches where 95% of all measurements of real pennies should fall.

Ask students to discuss their answers to Question 2b and to compare their thinking now with their conclusions in Question 2 of *Penny Weight*. Specifically, ask how helpful standard deviation was in making a decision. They may say the decision was just as easy to make without the use of standard deviation but that they are more confident or more knowledgeable about their decision when they use standard deviation.

The reasoning students are using here is at the heart of inferential statistics. It goes something like this: Given what we know about real pennies, how likely is it that this suspect penny is also real? To make this inference, the sample of pennies is treated as an approximation of the collection of all possible pennies. In this case, statisticians use a slightly different calculation of standard deviation. Students will be confronted with the following two different ways to calculate standard deviation when they learn to use their calculators to do this. You may want to discuss population standard deviation versus sample standard deviation with them now.

Standard Deviation: Population (σ) or Sample (s)

The computation students did for *Penny Weight Revisited* gave them the standard deviation of their sample data. However, in evaluating the authenticity of the supposedly counterfeit penny, what they really should consider is the standard deviation of the set of weights of all pennies in the world. If students respond that this is impossible, ask what they suggest instead. Try to draw out the idea of estimating the standard deviation of a population from a sample.

Technically, the best estimate of the standard deviation of a population, based on data from a sample, is given by a slightly different calculation than the one they used. Specifically, rather than divide the sum of the squares by n (the number of data items), one divides by $n - 1$, a smaller number. The statistic calculated in this way is usually represented by the letter s and is called the **sample standard deviation**.

If n is reasonably big, σ and s will differ only slightly. Using s means dividing by a slightly smaller number, so that you get a slightly larger estimate for the standard deviation of the distribution. One way to make sense of this is to consider that the sample of data only approximates the population. This adds an extra source of variability to the data, variability due to sampling error. Thus we would expect a measure of the variability in the sample to be slightly larger than the variability in the population.

Key Question

Does 9.9 seem about right for the standard deviation?

Can Your Calculator Pass This Soft Drink Test?

Intent

Students apply their understanding of the normal distribution and standard deviation to situations involving variability.

Mathematics

This activity asks students to reason from the properties of a normal distribution. Given the mean and standard deviation of several sets of data, they will be asked to reason about the amount of data that reside in various regions under the normal curve.

Progression

Students will work on this activity individually and then share their results in a class discussion.

Approximate Time

5 minutes for introduction

25 minutes for activity (at home or in class)

20 minutes for discussion

Classroom Organization

Individuals, followed by whole-class discussion

Materials

Can Your Calculator Pass This Soft Drink Test? blackline master (transparency, optional)

Doing the Activity

Suggest that students draw normal distributions, and label the mean and two standard deviations on either side of the mean, for each situation. Then they can justify their thinking using their curves and the expected percentages. If students need a reminder about the percentages related to a normal distribution, refer them back to the *Standard Deviation Basics* reference pages.

Discussing and Debriefing the Activity

For Question 1, students should use the fact that in a normal distribution, approximately 68% of all results occur within one standard deviation of the mean. Therefore, about 68% of the bottles should pass inspection, while 32% will need to be removed for correction.

You might paraphrase Question 1 as follows: **Suppose a bottle is chosen at random before the quality-control inspection. What is the probability that it ought to be removed?** Students should recognize that this is essentially the same question they just answered and that the probability is .32.

To lead into Question 2, ask, **How is Question 2 different from Question 1?** Question 2 is set up in an opposite manner to Question 1. That is, Question 1 provides boundary or cutoff values and asks for a percentage, whereas Question 2 gives the percentage and asks for a cutoff value. Specifically, the manufacturer wants to set a cutoff time so that only 2.5% of calculators will need repair before that time. Students will probably use a diagram of the appropriate normal curve to illustrate what this cutoff time should be.

Question 2 is complicated by the fact that it involves only one "tail" of the normal distribution. The horizontal axis of the diagram below shows the mean as well as the values one and two standard deviations above and below the mean. The shaded portions represent results that are at least two standard deviations from the mean.

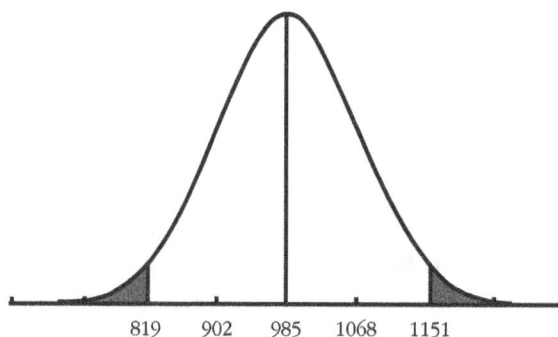

If students understand the percentages for standard deviation, they will see that the two shaded areas *together* total 5% of all calculators. Thus, the calculators that break down before 819 days constitute about 2.5% of the total. (Approximately 95% will break down somewhere between the day 819 and day 1151; another 2.5% will last more than 1151 days.)

In other words, if the manufacturer offers to replace calculators that break in less than 819 days, about 2.5% will need replacement.

Question 3 combines features of the first two questions. Like Question 1, it provides the cutoff value and asks for a percentage. Like Question 2, it is a "one-tail" problem, concerned only with the percentage of students whose scores are above a certain value. It is complicated by the fact that the difference between the cutoff and the mean is not a whole-number multiple of the standard deviation.

Again, a diagram is probably useful. In the diagram below, the shaded region is the area between the mean and one standard deviation above the mean. Students should be able to see that this area represents about 34% of all results. Thus, about 16% of students get scores above 610, and the percentage getting scores above 600 is slightly higher.

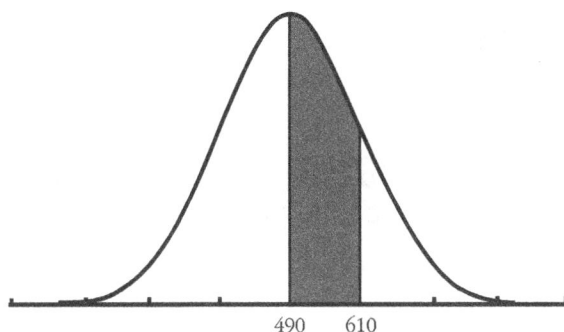

Don't get bogged down by the fine points of estimating the percentage of scores above 600. Although this is an interesting question, it is not the intended focus. You can mention that there are tables that provide details on areas like these so that people don't have to make visual estimates every time they come across a problem of this type.

Key Question

How is Question 2 different from Question 1?

Supplemental Activity

More About Soft Drinks, Calculators, and Tests (reinforcement or extension) involves an asymmetric "tail" problem (beyond two standard deviations at one end, beyond one standard deviation at the other end). Questions 2 and 3 require students to estimate areas under the normal curve whose boundaries are not whole multiples of the standard deviation from the mean.

A Standard Pendulum

Intent

In these activities, students return to the unit problem. They will use the tools they have developed to test ways to affect the period of a pendulum.

Mathematics

Earlier in the unit, students created a list of variables that might affect the time it takes a pendulum to swing back and forth one time. Now that they have developed an understanding of the **normal distribution** and such statistics as the **mean** and **standard deviation**, they will test whether changing the length of the pendulum, the weight of the bob, or the angle of the swing will lead to different periods. They will begin by gathering data using a standard "reference" pendulum. These data will provide the context for judging whether the results of a series of controlled experiments produce changes in the period that are greater than what one would expect from **measurement variation** alone.

Progression

Students begin by gathering and analyzing data from a standard pendulum. Then they conduct a series of experiments that isolate each of the three proposed variables to test whether any affect the period. In addition, students will revise and make presentations on the third POW and begin work on the fourth and final one of the unit.

The Standard Pendulum

Standard Pendulum Data and Decisions

Pendulum Variations

A Picture Is Worth a Thousand Words

Pendulum Conclusions

POW Revision

POW 12: Twelve Bags of Gold

The Standard Pendulum

Intent

Students return to the unit problem as they gather data about the period of a "standard" pendulum, which will be used as a point of reference for later experiments.

Mathematics

Armed with new tools—descriptive statistics like the **mean** and **standard deviation**, and the **normal distribution**—students investigate the behavior of a pendulum. Their goal is to determine which variables affect a pendulum's period and, eventually, to determine the nature of any relationships that exist. They will begin by analyzing the behavior of a reference pendulum with specific values for length, angle of swing, and weight.

Progression

Students will build a standard pendulum and then work in groups to collect several estimates of its period. The class will pool the data, compute the mean and standard deviation, and then create a normal distribution with these values.

Approximate Time

30 minutes

Classroom Organization

Groups, followed by whole-class discussion

Materials

Fishing line or dental floss

Weights (washers)

Scissors

Stopwatches

Doing the Activity

To get students to connect the concepts of normal distribution and standard deviation to the unit problem, you might have them look again at their posted plan of investigation and ask them to do focused free-writing on this topic: **Where are we now in the process of solving the unit problem?** Then ask them to share their ideas.

Say, **Now we're going to apply standard deviation to find out what determines the period of a pendulum.** Ask students how they might proceed in order to apply this tool to pendulum measurements and to complete step 1 in the plan of investigation—figuring out which variables affect a pendulum's period.

Bring out that they need to establish a point of reference. They can't talk about the period *changing* without having something from which it changes. In other words, the task is to establish a mean and standard deviation for a fixed, or standard, pendulum. Then they can investigate what happens to the period when one aspect of the pendulum is varied.

Because this idea is so central to the remainder of the unit, you might have several students address this question: **How would you state in your own words what we're going to do next?**

Now have students read the activity. The one-washer, 2-foot-long, 20° pendulum will be the reference pendulum; all others will be compared to it. Make sure the class understands that they will begin by doing some experiments using this standard pendulum, compiling their data, and calculating the mean and the standard deviation of the results of these experiments.

Once students have established the mean and the standard deviation, they will conduct experiments in which they will vary, one at a time, the weight, the length, and the amplitude to see whether these changes affect the period enough to be more than incidental measurement variation. These other experiments are part of the activity *Pendulum Variations*.

Ask, **How will you measure the period?** The class needs to agree on exactly how they will measure a single swing of the pendulum. In previous experiments, some students may have timed several swings and then divided by the number of swings, whereas others may have measured several individual swings and then averaged the results (see the discussion of *Building a Pendulum*).

If necessary, remind students of their experience with other experiments. For example, in *What's Your Stride?* they may have seen that it was more accurate to measure many strides and divide by the total number of strides than to measure individual strides.

The method to be used should be a class decision, arrived at through discussion and perhaps based on students' experiences with *Initial Experiments*. Record exactly what procedure students agree on, as they should use the same procedure throughout the rest of the unit. They may agree to time 5 swings and divide by 5. Or they may recognize that timing 10 swings and dividing by 10 can be done without a calculator and divides the measurement variation by 10 rather than 5.

Ask students to clarify exactly how they will take other measurements as well. For example, to determine the pendulum length, they might measure from where the string is attached at the top to the center of the washer. Theoretically, in a simple pendulum the weight at the end is treated as a single point, so it makes sense to measure to the center of the weight. Also emphasize that the amplitude is an angle measured *from the vertical*.

When the details have been established, have students work in their groups on the experiment described in the activity.

To get sufficient data to find reliable mean and standard deviation statistics, aim for a total of at least 30 trials. In a class of 32, with 8 groups each completing 5 experiments, the class will get 40 trials.

As groups complete each experiment, have them record each period on a master class list.

In the next activity, likely to be done as homework, students will work with this information, so they should copy the master list of times before leaving class.

Discussing and Debriefing the Activity

It is essential that everyone determines, and agrees on, the mean and standard deviation of the data from this standard pendulum. To do this, have students enter the data into their calculators and find both the mean and the sample standard deviation. Because the data here constitute a sample from "the world of measurements of the standard pendulum," it is more appropriate to use s, the sample standard deviation, than σ, the standard deviation calculation that treats the experimental data as the entire population.

You may want to go over the distinction between s and σ to clarify that s is an estimate of the standard deviation for the distribution of all measurements of the period. Also, be sure students realize that the mean of their data is only an estimate of the actual period of this pendulum.

On chart paper, sketch a normal distribution based on the class estimates for the mean and standard deviation. Save the graph for use in the activity, *Pendulum Variations*.

Key Questions

Where are we now in the process of solving the unit problem?

How would you state in your own words what we're going to do next?

How will you measure the period?

Standard Pendulum Data and Decisions

Intent

Students connect their work in *Statistics and the Pendulum* and their experiments in *The Standard Pendulum* in anticipation of the activity *Pendulum Variations*.

Mathematics

The theoretical **normal distribution** is a bell-shaped curve with a vertical line of symmetry at the mean, a section in the middle curved concave down, and tails curved concave up. The transition points in concavity—the inflection points of the graph—are the locations of one standard deviation above and below the mean. Experimental data derived from repeated measurements of some quantity will reflect measurement variation, and as a result the measured values will be approximately normally distributed. The mean and standard deviation of these data can be computed.

The class has sketched a normal distribution with the mean and standard deviation computed for their standard pendulum data. Now students will construct a frequency bar graph of the actual data, overlay a bell-shaped curve that approximates the distribution, and visually note the approximate locations of the mean and a standard deviation above and below the mean.

The final question in the activity—*How different must an observation be before you can be confident it isn't just measurement variation?*—is the key to determining whether changing a variable has a real effect on the period.

Progression

Students work on the activity individually and then share their work, including frequency bar graphs of their standard pendulum data, with the class.

Approximate Time

25 minutes for activity (at home or in class)

10 minutes for discussion

Classroom Organization

Individuals, followed by whole-class discussion

Materials

Class data from the standard pendulum experiments

Doing the Activity

This activity requires little or no introduction.

Discussing and Debriefing the Activity

Question 1 provides an opportunity to review ideas about measurement variation.

Ask for comments about the frequency bar graphs students created. If possible, get confirmation of the idea that the data items are approximately normally distributed and review the "normality assumption" stated in *Standard Deviation Basics*.

Have volunteers share their frequency bar graphs and accompanying normal curves, and discuss how they used these items to estimate the mean and standard deviation. This discussion is also an opportunity to review the ideas in the section "Geometric Interpretation of Standard Deviation" in *Standard Deviation Basics*.

The issue in Question 5 is best addressed by using the normal distribution graph with the mean and two standard deviations in either direction marked on the horizontal axis.

Pendulum Variations

Intent

Students perform experiments to test whether each of three variables affects the period of a pendulum.

Mathematics

Students will be conducting a series of controlled experiments in which they hold all of the pendulum variables equal to the values in the standard pendulum except for the one they wish to test. Then they examine the collected period data to see whether they fall within what one would expect due to measurement variation alone.

In physics, a simple pendulum is, in theory, one with a weightless string and a weighted bob that is a single point that oscillates through a small angle. With these simplifying assumptions, the period is a function of only the length of the pendulum. Practically speaking, students' pendulums only approximate a simple pendulum, so students might observe some additional variability.

Progression

Students work in groups to test each of the three variables in turn. Then the class draws conclusions about each variable.

Approximate Time

35 minutes for activity

30 minutes for discussion

Classroom Organization

Groups, followed by whole-class discussion

Materials

Additional washers and fishing line or dental floss

Class graph from *The Standard Pendulum*

Doing the Activity

Introduce the activity by referring to the list of variables that might affect the period of a pendulum. Then tell students that their work will now focus on these three variables:

- Weight of the bob
- Length of the string
- Amplitude

Then have the class read the activity.

Students are told to take each measurement twice but not to average their measurements. The reason for this is that the standard deviation for a set of *averages* of data is different from the standard deviation for the data set itself. Because the sample standard deviation found in *A Standard Pendulum* was from individual experiments, students need to look here at results from individual experiments.

Discussing and Debriefing the Activity

Focusing on one variable at a time, have groups share their measurements. Mark each measurement on the graph of the normal distribution of measurements of the period of the standard pendulum. Use a different color for each variable.

As each variable is discussed, students should examine their collective results and determine if that variable had a significant effect on the period based on whether there are data points that are more than two standard deviations from the mean.

Does weight matter? Students should see that their results for the pendulum with a change in weight seem to fit well within the results that would be expected for the standard pendulum.

Does length matter? Students should see that this variable change has dramatically affected the period; the results fulfill the criterion for "mattering." Bring out that the decision that "length matters" is based on the fact that the results would be extremely unlikely with the standard pendulum. Circle "length" on the list of possible factors.

Does amplitude matter? The variable of amplitude poses the most difficult "Does it matter?" question. An analysis based on the relevant laws of physics shows that changes in amplitude do affect the period of a pendulum, but the effect is very slight unless the angle exceeds 30° or so. For an increase from 20° to 30°, the change will probably be small enough to produce results within the measurement variation found for the standard pendulum. However, an increase from the standard 20° angle to a 60° angle might appear significant.

If the class experiments seem to show that amplitude does make a difference, students will have to make an assumption about amplitude in order to solve the original unit question.

If the class does not find a difference in pendulum period based on a change in amplitude, they will conclude that the period is determined only by a pendulum's length. You can mention that, in fact, very large changes in amplitude do affect the period, but that students can make the simplifying assumption described above.

It should be clear that the weight of the bob does not affect the period, that the length of the string does affect the period, and that the amplitude could, but probably doesn't, have a significant effect.

Tell students that for the rest of the unit, they will assume that a pendulum's length determines its period. You may want to return to the story to note that Poe's pendulum is "some thirty or forty feet overhead" and tell students that they will use 30 feet as the length of Poe's pendulum.

Ask students where they are on the outline of steps for solving the unit problem. They should see that they have finally finished step 1 and are now ready for step 2.

Key Questions

Does weight matter?

Does length matter?

Does amplitude matter?

What's next in solving the unit problem?

Supplemental Activity

Are You Ambidextrous? (reinforcement or extension) asks students to conduct a data-collecting experiment and then use standard deviation to decide whether a difference in results is significant.

A Picture Is Worth a Thousand Words

Intent

This activity contributes to the focus of the unit by examining how data can be used, or misused, to make a point.

Mathematics

Students consider what makes a graph effective in producing a given impact. They examine the same data plotted on graphs with differing scales, the visual effect of which is to communicate very different messages. They will create their own misleading graphs and look for such graphs in the media.

Progression

Students will work on the activity individually and then share their work in their groups and with the class.

Approximate Time

10 minutes for introduction

30 minutes for activity (at home or in class)

10 minutes for discussion

Classroom Organization

Individuals, then groups, followed by whole-class discussion

Doing the Activity

Have students take turns reading the activity aloud. You may want to have groups brainstorm topics to make misleading graphs about.

Discussing and Debriefing the Activity

Have students share their graphs in their groups. They should discuss what factors make graphs effective in producing a desired impact. Groups can then report to the class on the subjects of their graphs, as well as on one or two factors the group considers effective in producing startling effects.

Key Questions

What makes a graph more effective in producing a particular impact?

Do you think it is ethical to use graphs like these?

Pendulum Conclusions

Intent

In this activity, students write up and justify their conclusions from *Pendulum Variations* using the concepts that have been developed in the unit. They are then asked to grade the work of the members of their groups during the activities in *A Standard Pendulum*.

Mathematics

Using the language and ideas developed in the activities in *Statistics and the Pendulum* and *A Standard Pendulum,* students will express in writing the idea that changing the length of a pendulum is the only way to change the pendulum's period, beyond the changes one might expect from measurement variation.

Progression

Students will do this reflective work individually and share their conclusions with their group members and the whole class.

Approximate Time

10 minutes for introduction

25 minutes for activity (at home or in class)

15 minutes for discussion

Classroom Organization

Individuals, then groups, followed by whole-class discussion

Doing the Activity

Have students read Question 1, which asks them to summarize in their own words what they have learned from the experiments of the last few activities, using specific statistical language. You will also want to discuss Question 2, which asks them to grade their fellow group members on their work during the experiments of the last few days.

The conclusion of several days of group work offers a good opportunity for students to reflect on their ability to work collaboratively. Whatever grading criteria students use, ask them to provide evidence to support their evaluations. You may also want to assign your own grades evaluating each student's group-work skills.

Students may want to know whether their group members will see the grades they assign. Some teachers think students will be more honest in evaluating their peers if the evaluations are not shared. Others believe it's important for this information to be shared.

Discussing and Debriefing the Activity

You may want to let students read one another's summaries of conclusions from *Pendulum Variations* and discuss their reasoning in their groups. This discussion should reinforce the class's conclusion that length is the major factor in determining the period of a pendulum.

The follow-up of the peer grading might depend on whether students will see one another's evaluations.

- If they see the evaluations, this information can form the basis for a constructive group discussion about how to be a good group member.

- If they don't see the evaluations, the class can do focused free-writing on the topic "What my perfect group would be like." Afterward, you can ask volunteers to share their ideas.

In either case, you can also lead a discussion on what students have learned about the components of successful group work. The discussion should lead the class toward the realization that cooperation and participation of all group members are essential for group success.

POW Revision

Intent

There are at least two purposes for having students share their POWs. First, they get to see other students' work—both good and poor. Second, they receive feedback so that they can improve their own POWs.

Mathematics

This activity is designed to communicate the characteristics of high-quality work. It asks students to think about students' writing about a problem, including their own, rather than the problem itself. Thinking about thinking in this way is called *metacognition*. Research suggests that good problem solvers do a great deal of this kind of thinking and that students who practice metacognition improve their problem-solving skills.

Progression

Students will read and review the POWs of other students in class and then revise their own work outside of class.

Approximate Time

35 minutes for activity (in class)

30 minutes (at home)

Classroom Organization

Whole class

Materials

First drafts of *POW 12: Eight Bags of Gold*

Doing the Activity

Here are a few ways to have students review the POWs.

- Collect them and then pass them back in a different order.

- Have group 1 pass to group 2, group 2 to group 3, and so on.

- Make a chain by having each student pass to the person on the right.

If there are students who did not bring their POWs to class, you might have them read someone's POW after one reviewer has finished with it and is writing the review.

Suggest that in their reviews, students focus on two things: what they like about the write-up and how the write-up can be improved.

Be sure students understand their options for reworking their write-ups.

POW 12: Twelve Bags of Gold

Intent

This last POW of the unit is a more complicated version of *POW 11: Eight Bags of Gold*.

Mathematics

In this variation of *Eight Bags of Gold,* not only are there more bags, but we do not know whether the counterfeit bag is lighter or heavier than the rest. Students will be able to find the counterfeit bag without too much difficulty if they are not limited to three weighings. When the weighings are limited, the problem becomes much more difficult. Very few students have been able to solve it with three weighings. Many teachers have been unable to do so as well.

Progression

Students will begin work on the POW now, with presentations scheduled for late in the unit.

Approximate Time

30 minutes for introduction

1 to 3 hours for activity (at home)

20 minutes for presentations

Classroom Organization

Individuals

Doing the Activity

Tell students that this is an extension of *POW 11: Eight Bags of Gold* but much more difficult. Have students read over the activity and then ask, **How are *Twelve Bags of Gold* and *Eight Bags of Gold* similar? How are they different?**

Then give students class time to begin exploring this significantly difficult problem.

Discussing and Debriefing the Activity

After the POW presentations, ask whether other students had different ways of approaching the problem or arrived at a different solution, and have them share their ideas.

If no one was able to solve the problem with only three weighings, acknowledge the difficulty of the problem and perhaps suggest that students keep working on it.

Ask, **How was your work on *POW 11: Eight Bags of Gold* helpful (or might be helpful in the future, if you haven't yet solved the problem)?**

Graphs and Equations

Intent

The activities in *Graphs and Equations* use a series of explorations of graphs, In-Out tables, and equations to support the focus in the final set of activities, in *Measuring and Predicting,* on solving the unit problem.

Mathematics

In these activities, students will build on their work with patterns, functions, and their representations from *Patterns* and *The Overland Trail*. They will move among In-Out tables, graphs, and equations for several types of functions, including nonlinear functions. They will also explore ways to vary the shape of a graph by altering coefficients in an equation.

Progression

Graphs and Equations begins with several activities that focus on fitting functions to given sets of data points. Then it moves to an open-ended, technology-assisted exploration of various types of graphs. Finally, several activities ask students to find equations for given linear and nonlinear graphs.

Maliana the Market Analyst

Birdhouses

So Little Data, So Many Rules

Graphing Free-for-All

Graphs in Search of Equations I

Graphs in Search of Equations II

Graphing Summary

Maliana the Market Analyst

Intent

In this activity, the first of a series of activities that support the curve-fitting students will be doing to predict the period of a 30-foot pendulum, students sketch graphs from In-Out tables.

Mathematics

The data sets in this activity are nonlinear, so the graphs that students create—once they have determined the proper scales for the axes—will contain data points that they will then connect with smooth curves. Each of the data sets will have a turning point at which the graph changes from increasing to decreasing. However, the turning point might not be right at one of the data points in the table, so students will be required to guess the location of the maximum values of these nonlinear functions.

Progression

Students will create and analyze two graphs on their own and then share their results in their groups and with the whole class.

Approximate Time

20 minutes for activity (at home or in class)

15 minutes for discussion

Classroom Organization

Individual, then groups, followed by whole-class discussion

Doing the Activity

Before students begin, it may be helpful to discuss how they might choose scales that will allow them to see the graphs clearly.

Discussing and Debriefing the Activity

Have students compare their answers and share how they decided on the price that would maximize profit. If anyone connected the dots with line segments rather than with a smooth curve, ask whether prices between those given in the tables would necessarily have profits on those line segments.

Students should recognize that they cannot tell for sure what price would give the maximum profit. If they just choose the price among those listed that maximizes profit—$18 for a CD, $40 for a CD player—ask whether a price slightly higher or lower might result in a greater profit. (There is no definitive answer to this question.)

Students don't need to formulate equations to describe the data here; that idea will be pursued later in the unit.

Birdhouses

Intent

This activity is the second in the series in which students explore nonlinear relationships.

Mathematics

This activity asks students to make a prediction from two given data points. The most common approach, given the work students did in *The Overland Trail,* is to assume that the two points determine a linear function. In reality, an infinite number of functions can fit any two data points. Once a third data point is given, however, the nonlinear nature of the data becomes apparent. Even then, an infinite number of functions can be found that fit the three data points. Students are not expected to find an algebraic rule that fits the data in this activity, but they should be able to use graphs and tables to represent possible relationships.

Progression

Students will work on this activity in stages, with group work and whole-class discussion at each stage.

Approximate Time

30 minutes

Classroom Organization

Groups and whole-class discussion

Doing the Activity

Let students work for a few minutes on Questions 1 and 2, and then have them share their conclusions with the class. They will most likely have decided on an answer of 16 birdhouses for the 8 hours for Question 1 and concluded that the number of birdhouses that can be painted is double the number of hours.

Then have them continue with Question 3, emphasizing that you want them to think about *ways to approach the task* and that you don't expect them to find an equation that represents the data. Give them about 10 minutes to work on this question.

Discussing and Debriefing the Activity

Ask groups to report on what they did. Here are some approaches they may have tried.

- They may have applied ideas from *The Overland Trail* about approximating data with a straight line and looked for a linear graph or equation that would come close to fitting the given information.

- They may have tried to come up with a nonlinear function that would go through the three points (1, 2), (3, 6), and (5, 9). However, it is unlikely that anyone found a function that fits the data perfectly.

- They may have plotted the points and sketched a curve through them.

- They may have created a table and looked for patterns in it.

As you discuss the ideas, emphasize two key points: (1) that there may be more than one good answer and (2) even if they didn't find an exact answer, looking for a different kind of rule is a good approach.

Point out that when there were only two pieces of information, other rules would have fit them. The rule $y = (x - 1)^2 + 2$ is one nonlinear example that fits the two initial pieces of information.

The same principle applies for the three points. Just because a rule fits the data doesn't mean it must be *the* rule. People naturally think of the "two birdhouses per hour" rule for various reasons, including these: (1) It makes intuitive sense that Mia and her classmates paint the same number of birdhouses every hour. (2) That is the simplest rule that fits the information.

So Little Data, So Many Rules

Intent

This activity offers students opportunities to find several rules that fit a single data point.

Mathematics

An infinite number of functions can be found to fit a single data point. Their behavior—aside from what happens at that point—will vary widely. This activity gives students some experience with the diversity of functions at their disposal to address such a situation. As part of the discussion, they will be introduced to function notation.

Progression

Students will work on the activity individually and then share their results with their groups and the whole class.

Approximate Time

5 minutes for introduction

25 minutes for activity (at home or in class)

15 minutes for discussion

Classroom Organization

Individuals, then groups, followed by whole-class discussion

Doing the Activity

If time permits, present an example of a single data point for which students are to write a rule. For example, for the pair *In* = 1, *Out* = 2, three possible rules are $y = x + 1$, $y = x^2 + 1$, and $y = 3x - 1$.

Discussing and Debriefing the Activity

For each of the two questions, you might compile a list of all the rules that students created to fit the given number pair and then plot the number pairs for at least three of those rules.

Then ask, **What do your graphs from Question 1c have in common?** It may be obvious, but be sure students realize that any correct graph must go through the point (2, 5). Because their rules were created to fit the In-Out pair (2, 5), the graphs for those rules will also include the point (2, 5).

In preparation for later work in this unit, it is important that students see both linear and nonlinear examples for each case.

For a nonlinear rule, try to draw out the rule $y = \sqrt{x}$ for Question 2, as a square-root function will be needed for the final pendulum analysis. (Students will become aware of square-root functions in other ways before that time, so there is no need to push this topic if it doesn't come easily.)

Now tell students that they will learn another way of expressing a function, called *function notation.* Explain that, with function notation, we give each function a name consisting of a single letter. For instance, suppose that one of the rules found for Question 1 was $y = 2x + 1$ and that a student found the additional ordered pairs for this rule shown below.

In	Out
2	5
4	9
0	1
-2	-3

Ask for a letter of the alphabet, and tell students that we could use that letter as the name of this function. For this discussion, we will use the letter b.

Tell students that they can think of the table as "the b rule" and of the output as the result of "doing b" to the input. Explain that we write "doing b" to something, say x, as $b(x)$, and that the standard way to read this expression is "b of x." Some alternative, and perhaps clearer, ways to read this include "the b-value of x" and "b applied to x." You may want to use such phrasing initially, but students should gradually get used to the standard terminology.

Illustrate this new notation with examples. For instance, in the first row of the table, 2 is the *In* and 5 is the *Out*. Ask how this might be expressed using b. Help students see that they can express this using the equation $b(2) = 5$. It is standard terminology to refer to 5 as "the value of the function b at $x = 2$" as well as calling it simply "b of 2."

Ask how this notation could be used to write the general equation for the rule. Bring out that, instead of writing $y = 2x + 1$, the equation for the function can be written as $b(x) = 2x + 1$.

Have students practice this notation and language. Ask, for example, **What is $b(0)$? $b(7)$?**

Use similar examples to develop the insight that any letter can be used as the *In* to define the function. For instance, the equation $b(t) = 2t + 1$ defines the same function as the equation $b(x) = 2x + 1$.

Let students make up some letter combinations to stand for other functions that fit the given data pairs. For example, they might write $h(x) = x + 3$ or $f(x) = 5$ for Question 1, and $c(x) = x - 2$ or $g(x) = \sqrt{x}$ for Question 2.

This new way to represent functions symbolically is a good occasion to review the four approaches to functions that students have seen, especially as a lead-in to connecting functions to the unit problem.

- as a table of values (an In-Out table)
- as a graph
- symbolically (as an equation or rule or, now, using function notation)
- in terms of a situation

Ask, **Can you think of a way to express the goal of the unit in terms of function language and notation?** If necessary, have students express the goal in words first. They have seen that the period of a pendulum is determined primarily by the length of the pendulum, so they may say that they want to figure out what they can do computationally when they know the length in order to find the corresponding period.

If necessary, ask them to suggest variables to represent a pendulum's length and period and use them to express the goal using function notation. For example, if they choose L and P to represent the length and the period, they might express the unit goal like this.

Goal: To find an equation for a function f for which $P = f(L)$.

Ask more specifically what they want to know about this function in terms of the exact situation in Poe's story. If needed, remind them of their interest in the 30-foot pendulum. Help them to see that the goal for the specific situation can be expressed like this.

Goal: To find f(30), where f is the function that gives the period of a pendulum in terms of its length, which is 30 feet.

Students should see that if they have an equation or rule for $f(L)$, finding $f(30)$ is just a matter of substituting 30 for L.

Key Questions

What are all the rules you found for each number pair?

What do your graphs from Question 1c have in common?

How can you express the goal of the unit in terms of function notation?

Graphing Free-for-All

Intent

Students will experiment with "families" of functions and use graphing calculators to prepare presentations. The activity will strengthen their understanding of the connections among equations, tables, and graphs.

Mathematics

This activity is an informal introduction to transformations of functions. Students will begin to gain some experience with the graphical representations of function families like $y = mx + b$, $y = a(x - b)^2 + c$ and $y = a\sqrt{x - b} + c$. They will begin to build a set of notes that includes the various representations of each of these function families.

Progression

Students will work on this open-ended exploration in groups and then add to their own discoveries those of other students in the class.

Approximate Time

95 minutes for activity, presentations, and discussion

Classroom Organization

Groups and whole class

Materials

Poster-size graph paper

Doing the Activity

Introduce the activity and explain that, as part of groups' general exploration, they must explore the graphs of each of the following three functions and study variations on at least two of the three.

$$y = x \qquad\qquad y = x^2 \qquad\qquad y = \sqrt{x}$$

Discuss what you mean by "variations." For instance, variations on the function $y = x$ include $y = 5x$ and $y = x + 1$, and variations on $y = \sqrt{x}$ include $y = 2\sqrt{x}$, $y = \sqrt{5x}$, and $y = \sqrt{x+3}$.

Also discuss the write-up and presentation aspects of the activity. As groups work, assign specific functions or families for groups to prepare reports and posters for, which will help illustrate the general shapes of graphs in the various categories. Groups should prepare these reports and posters as they go, rather than waiting until they have completed the entire exploration.

Each poster should show an equation, its graph and the viewing rectangle used, and an In-Out table. You may want to urge students to use function notation as well as "$y =$" notation.

Discussing and Debriefing the Activity

As groups present, hang the posters to create a display of various functions and their graphs around the room.

You might have groups present their findings about their function families and then have the class enter certain equations in their calculators to view the families up close. Presenters should write down what should be entered into the calculators to create a given graph type. Be aware that students may use different viewing rectangles, in which case they will get different-looking graphs for the same equation.

Allow enough time after each presentation for audience members to add the sketch and other information to their notes. Each student can thereby create his or her own list of matching equations and graphs.

After all the function families have been introduced, focus on ways to think about the different families. Ask, **How can you organize all of these graphs in a systematic way?**

Have the class rearrange the posters, grouping similar graphs together. Let students decide what "similar" means, as well as what other criteria to use for this organizing process. It may be helpful to consider x-intercepts, y-intercepts, and shape.

Key Question

How can you organize all of these graphs in a systematic way?

Supplemental Activity

Family of Curves (reinforcement or extension) is a follow-up to this activity. Give each group the equation of a basic curve, such as $y = x^2$, and ask them to look at some simple changes that could be made to the equation, such as $y = x^2 + 2$ or $y = 3x^2$. Groups should then explore how graphs vary among members of that family of curves and make a poster showing their results. They can use graphing calculators or computers to assist with their work.

Graphs in Search of Equations I

Intent

This is the first of three activities in which students look for equations that fit given graphs.

Mathematics

Students are developing their skill at moving among the various representations of functions. Drawing on work in *The Overland Trail,* this activity asks students to find symbolic rules for three linear graphs. In the discussion, they will attend to the *x*- and *y*-intercepts of these graphs.

Progression

Students will work on this activity individually and discuss their ideas as a class.

Approximate Time

5 minutes for introduction

15 minutes for activity (at home or in class)

15 minutes for discussion

Classroom Organization

Individuals, followed by whole-class discussion

Doing the Activity

Graph a may need particular attention, because its equation does not involve *x*. An In-Out table should be especially helpful here, as students will see that the *Out* values are all equal to 5.

Discussing and Debriefing the Activity

During the discussion, illustrate the use of the terms **x-intercept** and **y-intercept**.

Ask what the intercepts are for each graph. Graph b has an *x*-intercept at ($\frac{2}{3}$, 0) and a *y*-intercept at (0, −2), though students might give a decimal estimate such as (0.7, 0) for the *x*-intercept. The only intercept for graph c is the origin (0, 0), which is both an *x*-intercept and a *y*-intercept. Graph a has a *y*-intercept, at (0, 5), and no x-intercept.

Let students know that an intercept is sometimes identified by a single coordinate. For instance, we might say that the y-intercept for graph b is −2. You might ask, **Why don't we need to state the x-coordinate for a y-intercept (or vice versa)?**

Enrich students' understanding by asking what the y-intercept means in a real-world context, such as those studied in *The Overland Trail*. They might recall that the *y*-intercept often gives the "starting value" for a situation. This view is particularly appropriate if the horizontal axis represents time.

Finally, you might also raise the question of how many intercepts of each kind a graph might have. Students may be able to explain why a graph of a function cannot have more than one
y-intercept and perhaps will be aware that it can have any number of *x*-intercepts.

Key Question
What are the *x*- and *y*-intercepts for each graph?

Graphs in Search of Equations II

Intent
Students continue finding equations for graphs—in this case, for nonlinear graphs.

Mathematics
This is the third of four activities designed to help build intuition about the graphs of functions with certain types of rules. Students are being introduced to certain families of elementary functions and their symbolic, graphical, and tabular representations.

In this activity, the focus is on quadratic functions of the form $y = a(x + b)^2 + c$ and square-root functions of the form $y = a\sqrt{x + b} + c$. The square-root function is presented graphically as a quadratic function rotated 90 degrees. This affords the opportunity to discuss that, in this case, x is a function of y.

Progression
Students will work on the activity individually and then share their results in class.

Approximate Time
15 minutes for activity (at home or in class)

15 minutes in class

Classroom Organization
Individuals, followed by whole-class discussion

Doing the Activity
As in *Graphs in Search of Equations I,* making In-Out tables for these graphs will help students look for the patterns that will, in turn, assist them in finding equations.

Discussing and Debriefing the Activity
Ask students to share both their answers and their solution methods.

Students will likely not have trouble with graphs a or b, but for graph b they should be careful not to place the negative sign inside parentheses. That is, they should not write $y = (-x)^2$.

You may want to remind students that graph c does not represent a function, because each positive x-value has two corresponding y-values.

In particular, be sure students see that the upper half of graph c is the graph of the equation $y = \sqrt{x}$. This example is especially important, because the data for the pendulum experiments relating period to length should more or less fit an equation of the form $y = c\sqrt{x}$ (but don't tell students this, as it is their job to figure out what sort of function describes a pendulum's period).

If students try to use a calculator to draw graph c, they will have to graph the functions $y = \sqrt{x}$ and $y = -\sqrt{x}$ simultaneously in the same window.

Graphing Summary

Intent

In this final activity in *Graphs and Equations,* students organize what they have learned over the past few days about equations and their graphs.

Mathematics

This is the last of four activities in which students build their intuition about different types of functions and their representations and transformations.

Progression

Students will begin this activity individually and then share and combine their summaries in a class discussion. They will include their work on this activity in their unit portfolios.

Approximate Time

30 minutes for activity (at home or in class)

25 minutes for discussion

Classroom Organization

Individuals, followed by whole-class discussion

Materials

Work on the activities *Graphing Free-for-All, Graphs in Search of Equations I,* and *Graphs in Search of Equations II*

Doing the Activity

Depending on how students took notes as groups presented their function families, this summary should reinforce all they have learned about intercepts and shapes of data.

Discussing and Debriefing the Activity

The time spent helping students to use everyone's work to build a set of useful notes—documenting different function types and their rules, In-Out tables, and graphs, along with ways to transform them—will pay dividends in solving the unit problem and in later units in the curriculum.

You might work as a class to make posters of each function family, including linear functions, square-root functions, quadratic functions, and exponential functions.

Measuring and Predicting

Intent

In these activities, students will solve the unit problem and build an experimental model to confirm their solution.

Mathematics

The mathematical focus of the activities in *Measuring and Predicting* is on collecting data relating the length of a pendulum to its period, fitting a function to those data, and then making a prediction for the time it takes a 30-foot pendulum to complete 12 swings. The relationship between the period and the length is a function of the form $y = c\sqrt{x}$.

Progression

Students will conduct experiments in groups, make predictions, and then test those predictions as a class by constructing a 30-foot pendulum. The unit concludes with presentations of the final POW and compilation of the unit portfolio.

An Important Function

Graphs in Search of Equations III

The Thirty-Foot Prediction

Mathematics and Science

Beginning Portfolios

The Pit and the Pendulum Portfolio

An Important Function

Intent
In this activity, students collect data about the periods of pendulums of varying lengths in preparation for predicting the period of a 30-foot pendulum.

Mathematics
In earlier investigations, using what they have learned about measurement variation, students narrowed down the possible variables that might affect the time of the swing of a simple pendulum. In this activity, they perform controlled experiments to collect data on the nature of the relationship between the length of a pendulum and its period.

Progression
Students work in groups to build pendulums of varying lengths and then conduct experiments to gather data. They compile their results in an In-Out table that the entire class will then use to solve the unit problem in the activity *The Thirty-Foot Pendulum*.

Approximate Time
25 minutes

Classroom Organization
Groups

Doing the Activity
Review the status of the class's progress on the unit problem. This review depends on the class analysis of the amplitude variable in the discussion of *Pendulum Variations*. However they resolved that issue, explain that today's experiments will treat length as the only variable.

With that in mind, discuss these questions:

How will you gather more data on length?

What factors will you hold constant so that only the length varies?

How will you make sure the data you collect are reliable?

As a class, decide on the pendulum lengths to be tested. Remind students that the more points they use and the wider the range of points, the less likely it will be that several different functions could explain their data.

Make sure to include both fairly long pendulums, including 6 feet and 8 feet, and fairly short ones, including 1 foot and 0.5 foot. Otherwise the collected data may appear to lie in a straight line. (Students probably shouldn't test beyond 10 feet or so, because the impact of being able to use the data to make a prediction will be more dramatic if the 30-foot pendulum is significantly beyond students' earlier results.)

With a class of 36 students, you can have each group of four investigate a different length, such as 0.5 foot, 1 foot, 2 feet, 3 feet, 4 feet, 5 feet, 6, feet, 7 feet, and 8 feet. Students may have to tape long pendulums from bookcases or the ceiling.

Remind groups that the activity asks them to find the time for 12 periods.

Have groups record their data in a single class In-Out table, such as the one shown below, with length as the In and time for 12 periods as the *Out*.

Pendulum length	Time for 12 periods

Discussing and Debriefing the Activity

When all the data are posted, ask the class, **What do you notice about the data?** Students are likely to notice that, as the length of the pendulum increases, so does the period.

Remind students that their goal is to figure out *from the data* how long it would take a 30-foot pendulum to make 12 swings. **Do you think the relationship between length and period is linear?** Someone might suggest graphing the data to find out.

You might also mention that after they have completed their analysis (in the activity *The Thirty-Foot Prediction*), they will test their ideas by actually building a 30-foot pendulum. You may want to spend some time talking about how they might do that, because it will happen soon.

Key Questions

How will you gather more data on length?

How will you make sure the data you collect are reliable?

What factors will you hold constant so that only the length varies?

What do you notice about the data?

Do you think the relationship between length and period is linear?

Supplemental Activity

More Height and Weight (extension) follows up on the supplemental activity *Height and Weight* by posing the question with a more specific focus. It also incorporates students' work with curve-fitting.

Graphs in Search of Equations III

Intent

This is the third in a series of individual activities in which students develop their skill at finding equations for graphs of different types.

Mathematics

Students will be searching for equations for transformed quadratic functions of the form $y = a(x + b)^2 + c$. They will work from graphs and In-Out tables as they search for patterns.

Progression

Students will work on this activity individually and then share their results in a class discussion.

Approximate Time

20 minutes for activity (at home or in class)

10 minutes for discussion

Classroom Organization

Individuals, followed by whole-class discussion

Doing the Activity

The graphs in this activity may be more challenging than any that students have tried to identify up to now. Their work from *Graphing Free-for-*All and *Graphing Summary* should be helpful to them.

Discussing and Debriefing the Activity

Ask students to share the strategies they used to find an equation for each graph. They can use graphing calculators to check their equations.

If they are having difficulty, ask what familiar graph these graphs resemble (namely, $y = x^2$) and how they differ from that graph.

Key Questions

How are these graphs similar to and different from those in Graphing Free-for-All?

In what family of functions do these graphs fit?

What are the coordinates of the vertices of these graphs?

The Thirty-Foot Prediction

Intent

In this, the unit's ultimate activity, students look for a function to fit their pendulum data and to enable them to predict the period of a 30-foot pendulum. Then they measure the time needed for a real 30-foot pendulum to make 12 swings and compare their experimental results with their predictions.

Mathematics

Students will begin by trying to fit a curve to the data they collected in *An Important Function*. They will use the function, *f*, that they find to compute *f*(30). They will then test this prediction by building and timing a 30-foot pendulum.

Progression

Students will work on the curve-fitting part of this activity in groups, share their predictions with the class, and then build and test a 30-foot pendulum as a class.

Approximate Time

40 minutes for activity

35 minutes for discussion

Classroom Organization

Groups and whole class

Materials

Class data table from *An Important Function*

Materials for building a 30-foot pendulum, such as nylon fishing line and a 5-pound weigh (Sufficient weight is needed to keep the line taut.)

Doing the Activity

Students have finished almost all the necessary preparation for answering the unit problem. They have found out which variable affects the period, and they have collected relevant data. They are now ready to look for a pattern in their data and use it to predict the time it will take for a 30-foot pendulum to make 12 swings.

Remind students of the curve-fitting they did using graphing calculators in *The Overland Trail,* and explain that they will use the same technique to analyze their pendulum data. You may want to review the general steps.

- Plot the data on the graphing calculator or in Fathom Dynamic Data™ software.

- Leave the data on the screen and graph a function that you think might approximate the data well.

- Examine how closely your function's graph approximates the data. Adjust the function until you think it approximates the data about as well as possible.

If groups need a hint as they look for a function, ask whether the graph resembles any of the examples from recent activities, or have them look at the posters from *Graphing Free-for-All* and *Graphing Summary*.

Ask students, **What function did you find to fit the data?** If they have accurate data with a wide enough range, with L representing the pendulum length and P representing the time for 12 periods, they should find that there is an equation of the form $P = c\sqrt{L}$, for some constant c, that closely fits the data. You may get several curves of this form. The class need not agree on which function best fits the data.

What is your prediction for the period of a 30-foot pendulum? Collect and post groups' values for $f(30)$ for later comparison with the results of the upcoming experiment.

Raise the question of whether extending the data in this way makes sense. **Do you think this is a reliable method for making a prediction? Do you have any reason to believe that the pattern of the data will continue as far out as 30 feet?** All you need to accomplish here is to raise some skepticism. Students' curve-fitting should work well in this situation, but they should be aware that this is a complex issue. The real test will come when they actually conduct the 30-foot pendulum experiment.

To construct the pendulum, you might hang it from the back of the bleachers in the school stadium or gymnasium or ask the fire department or a utility company to provide a truck from which the pendulum can be swung.

If you are unable to set up a 30-foot pendulum, build one that is as long as possible and have students predict the time it will take for a pendulum of that length to make 12 swings.

Have several timers time each swing, and average their results.

Discussing and Debriefing the Activity

Once the data have been collected, discuss whether the functions students found accurately predict the period of the actual 30-foot pendulum. If they do not, let students speculate why not.

Then turn back to the opening story and ask, **Do you think Poe's story is realistic? Do you think the prisoner could have escaped in the amount of time it takes a 30-foot pendulum to swing 12 times?**

To give students a better sense of how long that time is, ask them to close their eyes and try to estimate when a time interval has elapsed that is equal to the time they measured for the 12 swings of a 30-foot pendulum.

Key Questions

What function did you find to fit the data?

What is your prediction for the period of a 30-foot pendulum?

Do you think this is a reliable method for making a prediction?

Did your predictions match the actual period

Do you think the prisoner could have escaped in that amount of time?

Supplemental Activity

Out of Action (reinforcement or extension) asks students to fit a curve to data to make a prediction and then use that prediction to make a decision. This activity might be used late in the unit, as a follow-up to students' work on the unit problem.

Mathematics and Science

Intent

This activity gives students a chance to reflect on important ideas and concepts in the unit and how they might be useful in solving a problem.

Mathematics

Mathematics is a science in its own right, as well as an essential tool for scientists. For example, there is inevitable error in making physical measurements, and mathematics provides a way to describe and account for it. And mathematical objects (such as functions) can be used to model physical processes.

Progression

This activity will be done individually and then discussed briefly in class.

Approximate Time

20 minutes for activity (at home or in class)

10 minutes for discussion

Classroom Organization

Individuals, followed by brief whole-class discussion

Doing the Activity

Before students begin work, brainstorm with them the many mathematical and scientific ideas that were studied in this unit. Also mention that this activity will be included in the unit portfolio.

Discussing and Debriefing the Activity

Have volunteers share the problems they wrote about and how mathematics and science concepts they have learned about could be helpful in solving them.

Supplemental Activities

Gettin' On Down to One is a POW-like activity in which students investigate the behavior of number sequences and the relationships between that behavior and the number chosen to start a particular sequence.

Programming Down to One asks students to write a computer or calculator program to generate the number sequences in *Gettin' On Down to One*.

Beginning Portfolios

Intent

Students begin work on their unit portfolios by stating the unit problem in their own words and writing about how they arrived at a solution for that problem. In the process, they reflect on various aspects of what they have learned and prepare for the unit assessments.

Mathematics

This unit provides a good example of a problem in which science and mathematics support each other. In this unit, students designed scientific experiments and used mathematics to describe and account for the inevitable variation in the measurements they made. Now they begin to compile evidence of what they have come to understand about the solution to the unit problem.

Progression

Students work on the activity individually and then discuss it briefly in class.

Approximate Time

5 minutes for introduction

25 minutes for activity (at home or in class)

10 minutes for discussion

Classroom Organization

Individuals, followed by brief whole-class discussion

Doing the Activity

Prior to this beginning work on the unit portfolios, remind students to bring all their work from the unit to class. You might brainstorm with them how they arrived at a solution to the unit problem.

Discussing and Debriefing the Activity

Have volunteers share the key ideas and concepts they identified and how the activities they have engaged in relate to those ideas and concepts.

The Pit and the Pendulum Portfolio

Intent

Students review and document their mathematical activity and learning during the course of the unit. Their product is an opportunity for assessing what they have learned and what they believe is important in their learning (see "About Portfolios" in the *Overview to the Interactive Mathematics Project*).

Mathematics

The big mathematical ideas addressed in this unit are described in the introduction to *Pit and the Pendulum.*

Progression

Students are introduced to the particular expectations for *The Pit and the Pendulum* portfolio. They review their materials and begin to compile their portfolios.

Approximate Time

20 minutes for introduction

50 minutes for activity (at home or in class)

Classroom Organization

Whole-class introduction, then individuals

Doing the Activity

Have students read the instructions in the student book carefully.

If students do not complete the task in class, have them to take the materials home and finish compiling their portfolios for homework. They should bring the portfolio, with the cover letter as the first item, to the next class.

Discussing and Debriefing the Activity

You may want to have students share their portfolios in their groups, comparing ideas about the mathematical and science concepts they wrote about in their cover letters and the activities they selected.

What's Normal?

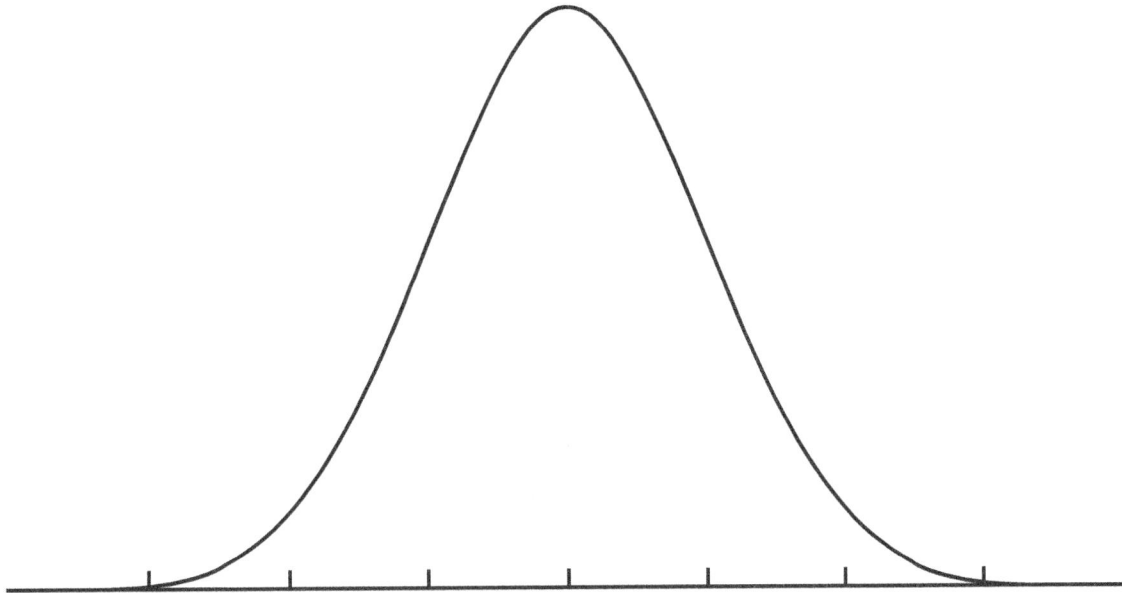

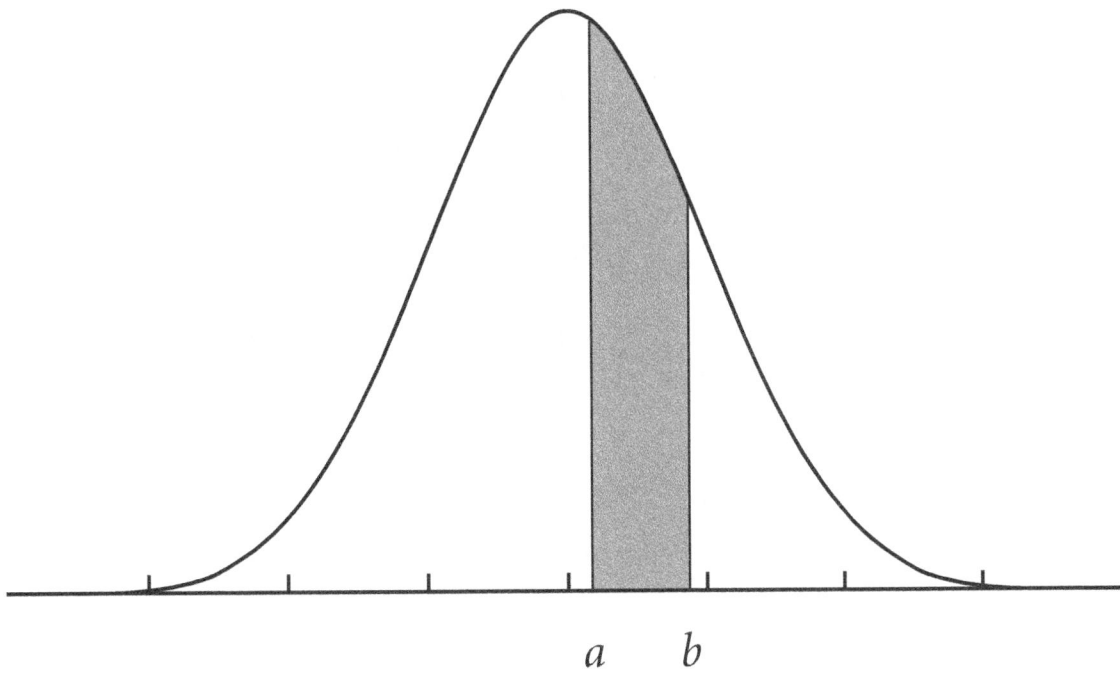

a b

What's Normal? (continued)

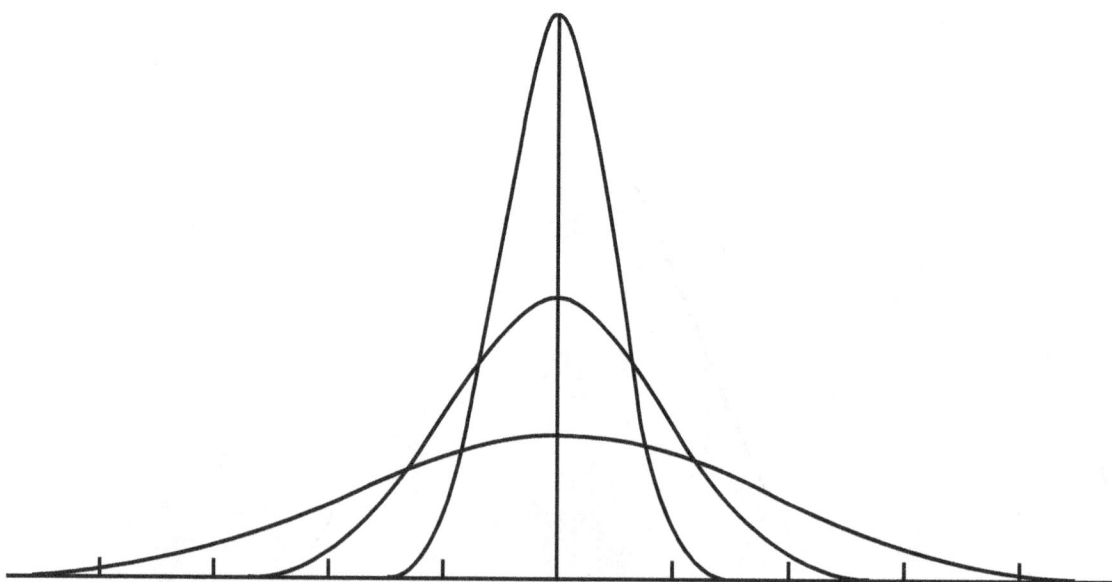

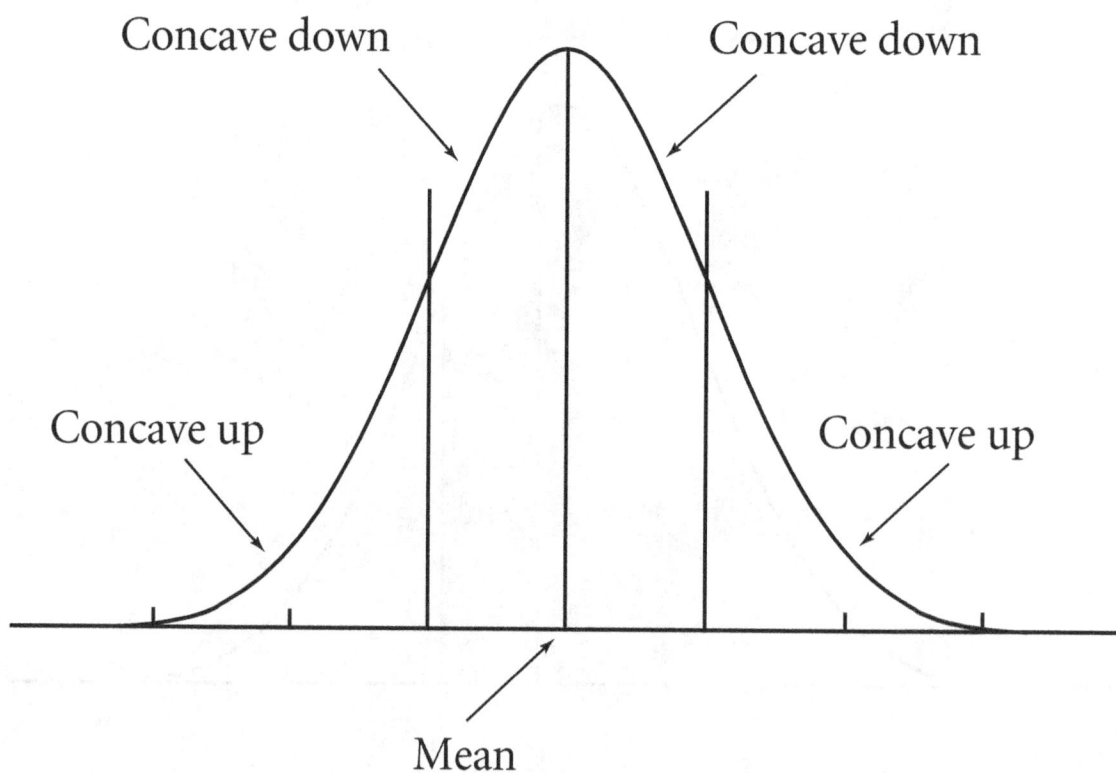

Concave down   Concave down

Concave up   Concave up

Mean

An (AB)Normal Rug

1.

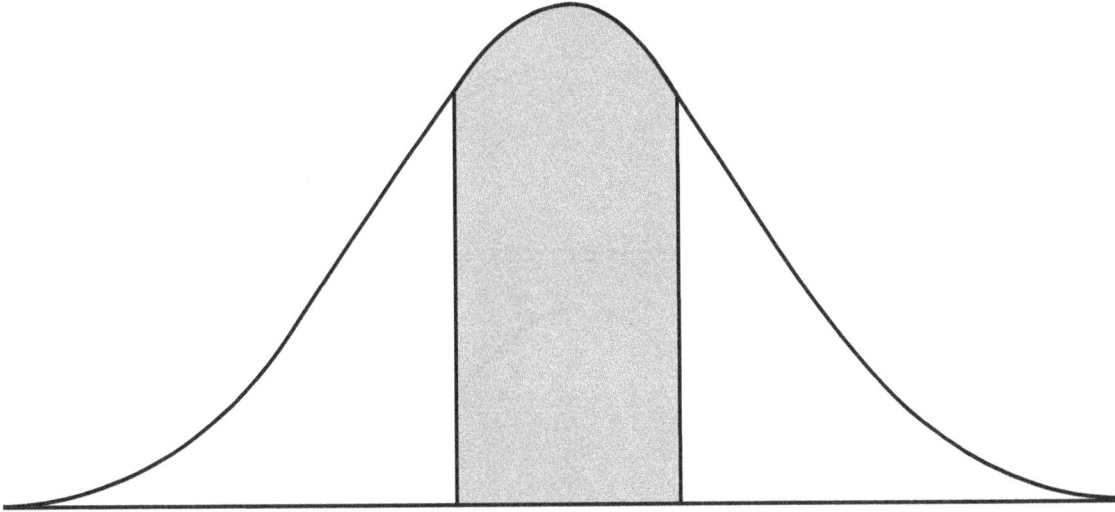

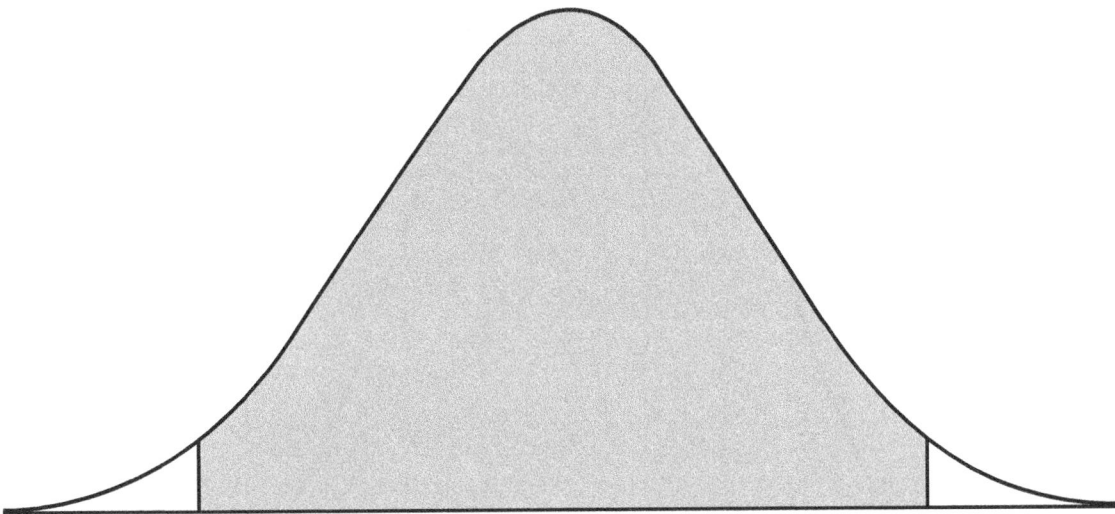

An (AB)Normal Rug (continued)

2. and **3.**

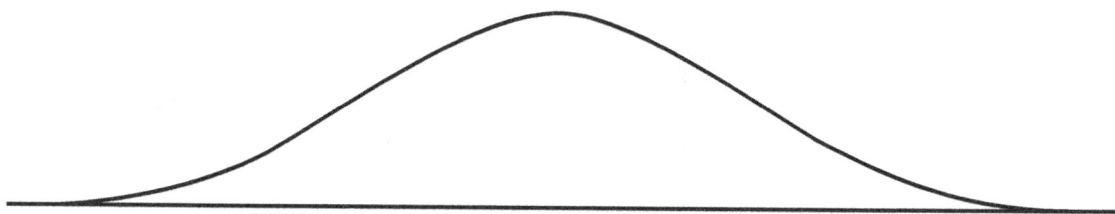

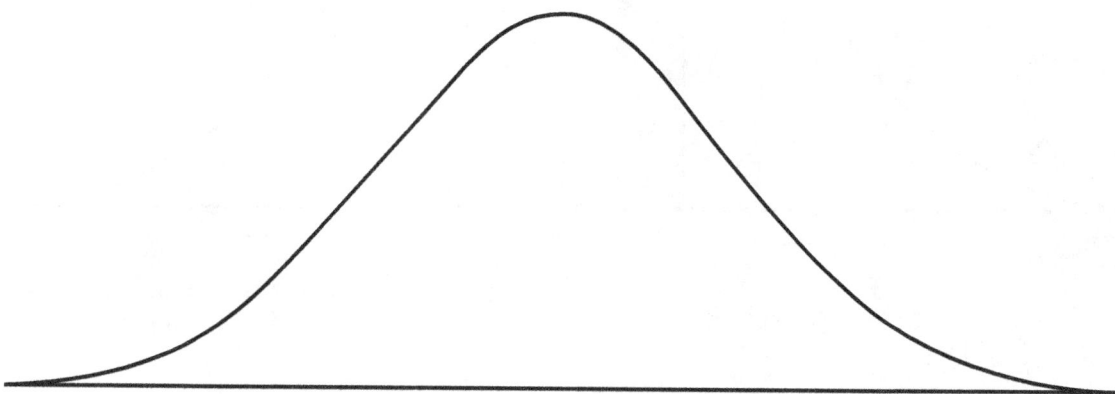

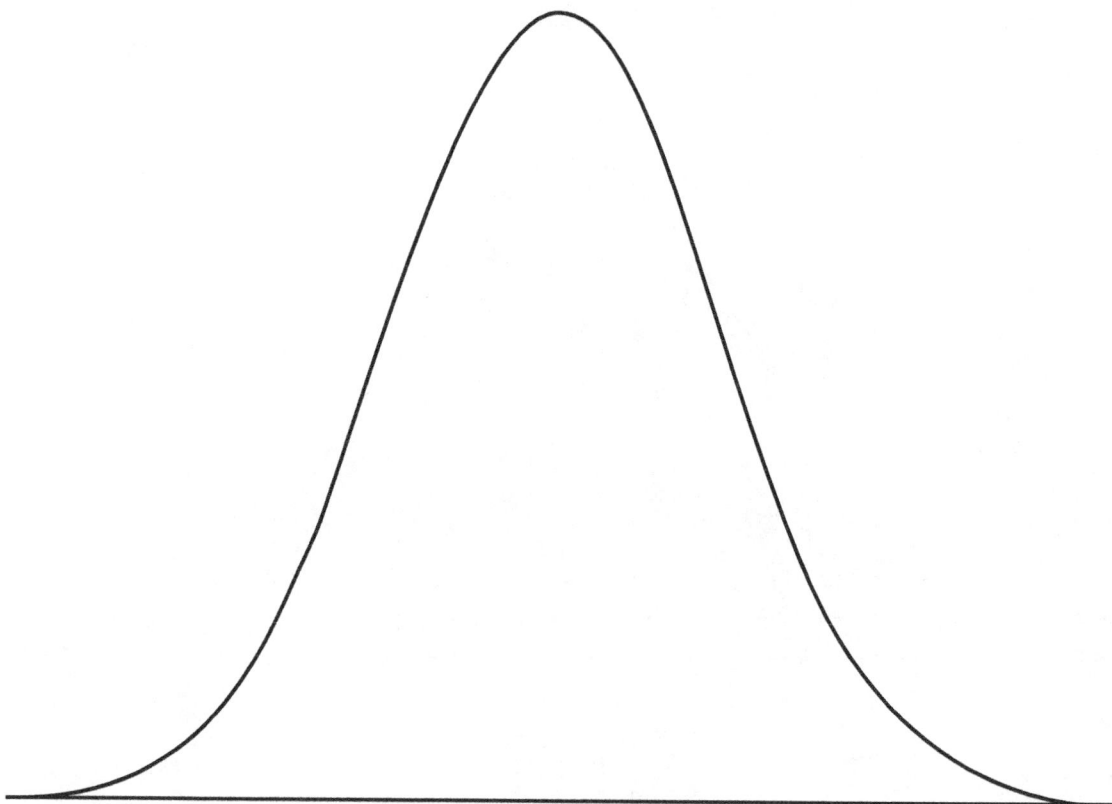

Standard Deviation Basics

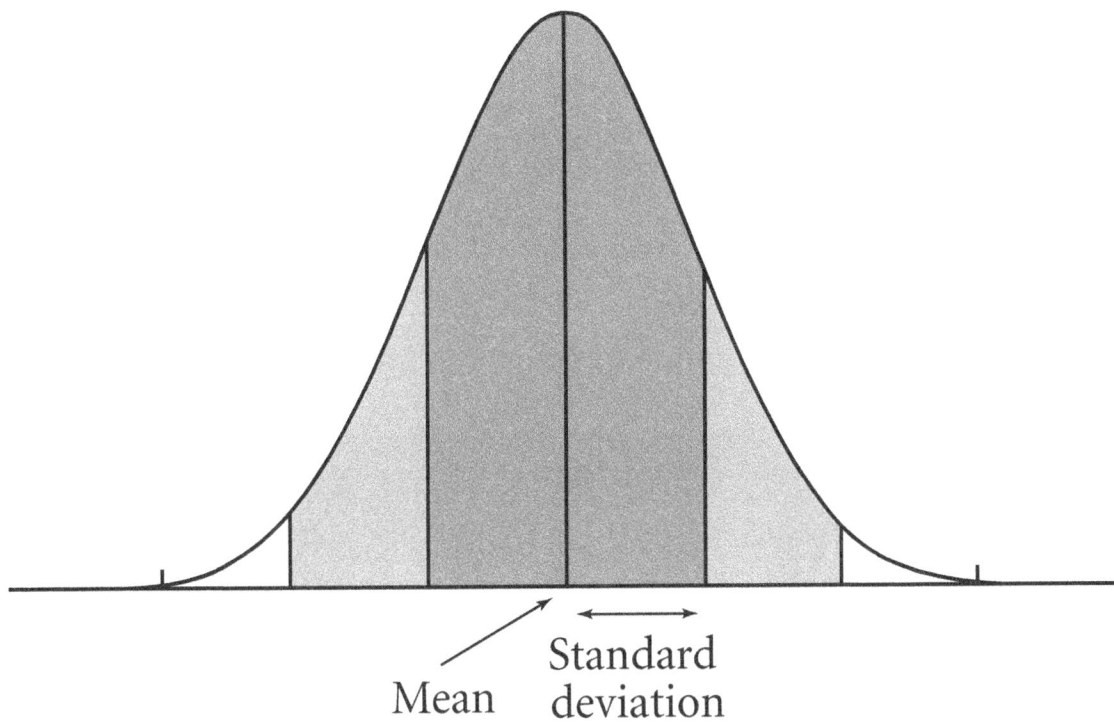

Mean

Standard deviation

Penny Weight Revisited

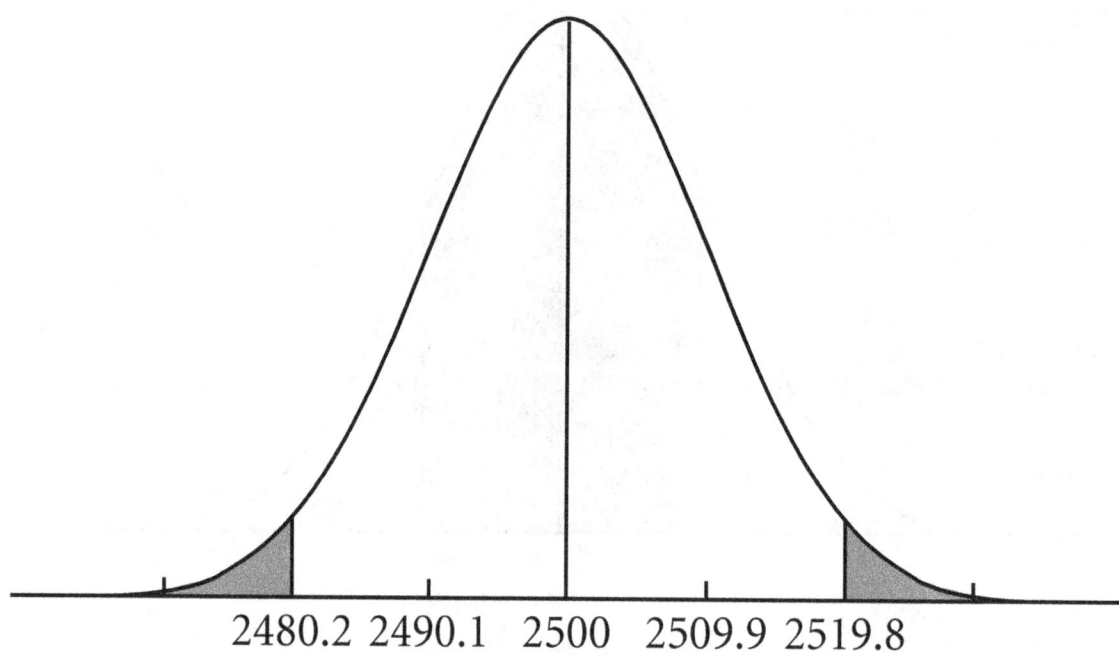

2480.2 2490.1 2500 2509.9 2519.8

Can Your Calculator Pass This Soft Drink Test?

2.

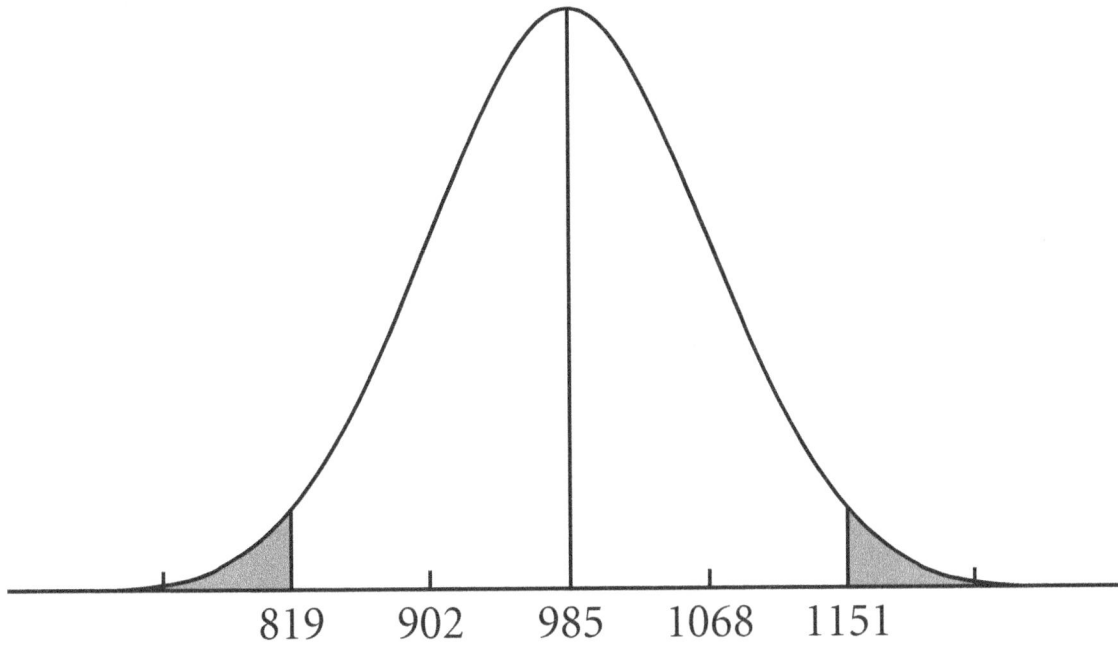

819   902   985   1068   1151

3.

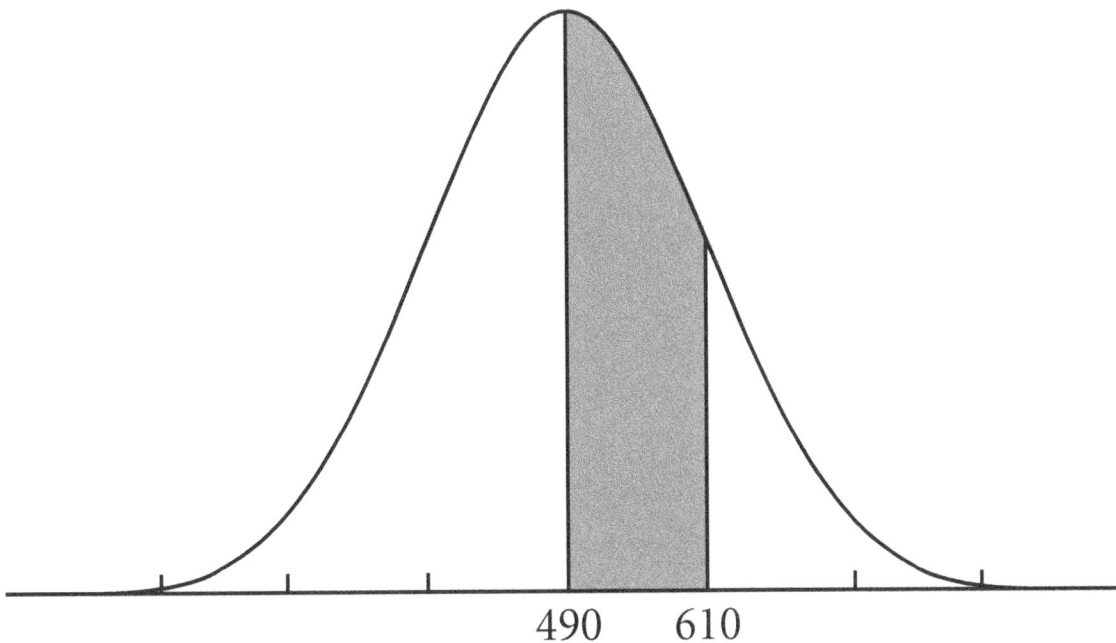

490   610

1-Inch Graph Paper

$\frac{1}{4}$-Inch Graph Paper

$\frac{1}{4}$-Inch Graph Paper

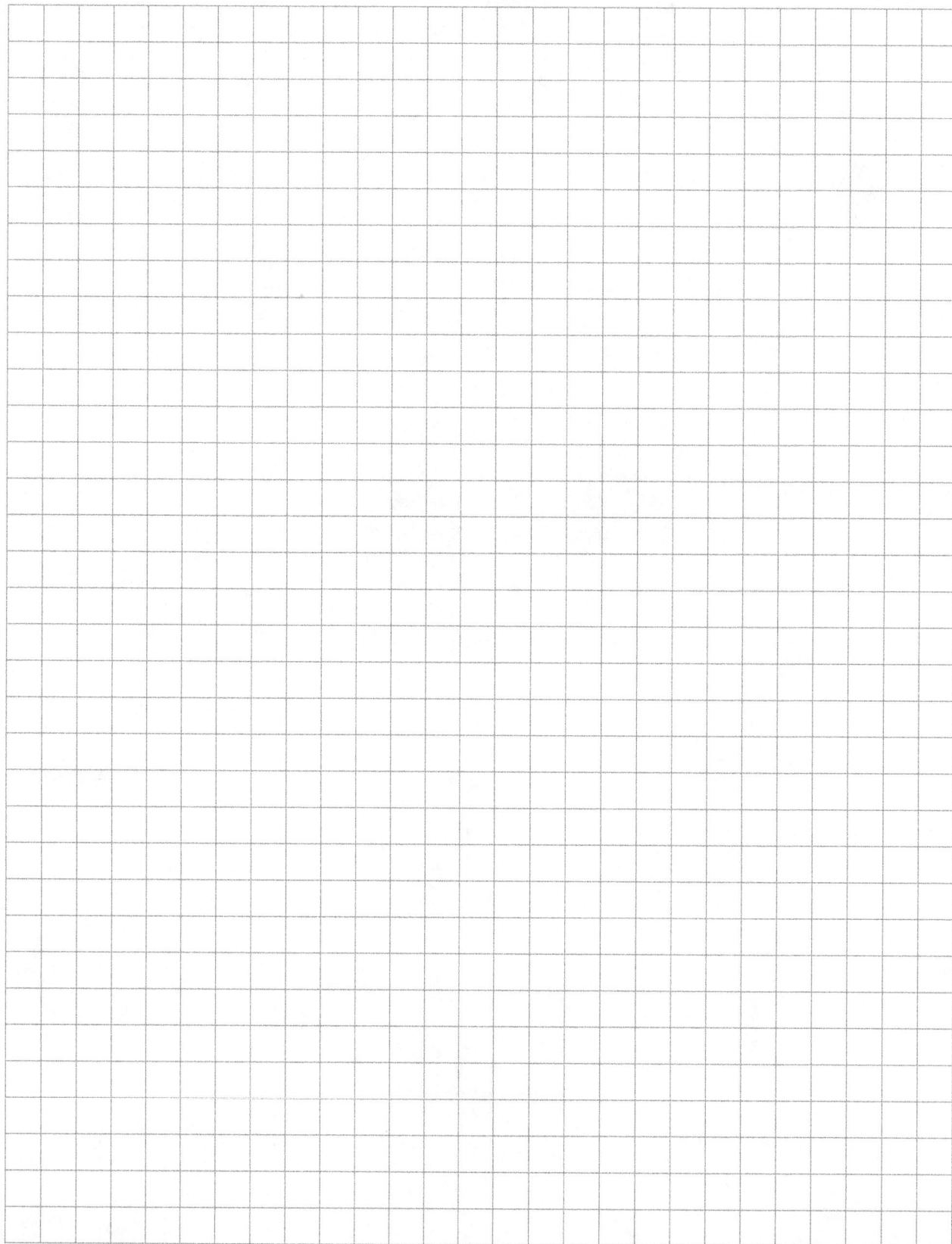

Steve works in a hobby shop. He was looking at the relationship between the length of certain models and the amount of paint they require.

Here are some estimates he came up with.

Length of model (in feet)	Amount of paint (in ounces)
1	2
2	6
3	14
5	40

Assume this pattern continues. Estimate how much paint would be needed for a model 10 feet long. Explain your reasoning.

Lana, a circus performer, does bicycle tricks. She wants to ride her bicycle right up to a brick wall and then stop dramatically. She wants to stop very close to the wall without crashing.

Lana needs to know when to apply the brakes. She doesn't want to try the experiment because she is afraid of crashing. She wants to predict ahead of time at what point she should hit the brakes. She also realizes that no matter how hard she tries to make the conditions the same every time, there may be some variation in the distance required to stop her bicycle.

Devise a plan to collect and analyze data that will allow Lana to make this prediction. Describe how she might use the data

In your answer, apply the ideas from the unit. Discuss the variables to consider and the problems Lana might encounter. Use the concepts of normal distribution and standard deviation in your answer.

The Pit and the Pendulum Guide for the TI-83/84 Family of Calculators

This guide gives suggestions for selected activities of the Year 1 unit *The Pit and the Pendulum.* The notes that you download contain specific calculator instructions that you might copy for your students. NOTE: If your students have the TI-Nspire handheld, they can attach the TI-84 Plus Keypad (from Texas Instruments) and use the calculator notes for the TI-83/84.

In this unit students learn a variety of statistical concepts and use them to interpret collections of data. Two of these concepts are the mean and standard deviation of a data set. These can be calculated easily and quickly on a graphing calculator. Once the data set is in the calculator's lists, you can plot it easily as a frequency bar graph. This unit also extends graphing ideas from previous units. Students explore different types of nonlinear functions in the activity *Graphing Free-for-All.* They fit a nonlinear function to the data set they collected from swinging pendulums of different lengths.

Because this unit contains many activities in which students represent data sets as frequency bar graphs, the "Making a Frequency Bar Graph" notes are a useful reference throughout. The notes require some basic familiarity with lists. You might refer back to "Using Lists to Build a Spreadsheet" in *Patterns Notes for the TI-83/84 Family of Calculators.*

You might show class data for various activities in a calculator with overhead viewing capabilities. (With an adapter, any TI-84 model calculator can connect to your ViewScreen.) For large data sets, use alternative technology, as the storage capacity of the TI-83/84 is 999 entries.

Time Is Relative: In this activity, students collect data and graph it using a frequency bar graph. You might enter the data on a calculator and have the calculator build the frequency bar graph, using the note "Making a Frequency Bar Graph." (These notes for *The Pit and the Pendulum* are similar to those in *The Game of Pig,* but they have been adapted to fit the context of this activity.) You will need to post a frequency bar graph of the class data from *Time Is Relative.* If you plan to use list L1 again for other data sets, move the data from *Time Is Relative* to another list. To move data from one list to another, place your cursor on the label of the new list (L6, for example), enter the name of the old list (L1, for example) and press ENTER . You can also adjust your **STAT PLOTS** menu so that either Plot 2 or Plot 3 refers to the new list. Turn off a plot when you don't want to view that particular graph.

What's Your Stride?: You might have students transfer or recreate the frequency bar graph of the data set on a calculator with overhead viewing capabilities. Keep in mind that you will want to save this graph for use later. To prevent the data set from being erased, you can move it to a different list. You can easily adjust the size of the bars in your graph on a calculator by pressing WINDOW and changing **Xscl**. For more information about creating frequency bar graphs, see the discussion in "Making a Frequency Bar Graph."

Pulse Analysis: In this activity, students calculate the means of several measurements, then use their data sets to build frequency bar graphs. They can do both of these tasks efficiently on a graphing calculator. A graphing calculator can find the mean of a data set in several ways. "Calculating the Mean and Median of a Data Set" (in *The Overland Trail Notes for the TI-83/84 Family of Calculators*) explains one way. *The Pit and the Pendulum* note "Calculating Statistical Measures" has directions for accessing the calculator's 1-variable statistics command, which includes the mean.

Flip Flip: Students can easily simulate the flip of ten coins with **randInt(0,1,10)**, where 0 = tails and 1= heads. Because each result is a list of zeros and ones, the number of heads in each trial is really the numerical sum of the list. So, if students press 2nd [**LIST**] and select **sum(** from the **MATH** submenu, they can create a slightly-more sophisticated simulation with **sum(randInt(0,1,10))**, which indicates the number of heads rather than the individual outcomes. By repeatedly pressing ENTER, you can simulate the ten-coin toss again and again. You may want to display the frequency bar graph of the classroom results on a calculator with overhead viewing capabilities. If any students have become proficient at programming, challenge them to write a program that repeats this simulation for a user-defined number of trials and stores the outcomes in a list; for an additional challenge, students can research how to write instructions for displaying a frequency bar graph as part of the program.

What's Rare?: The discussion of this activity relies on three other frequency bar graphs: the graph from *Time Is Relative*, the graph from *What's Your Stride?,* and the graph from *Pulse Analysis.* If you have saved these lists on your calculator, set up the **STAT PLOTS** menu appropriately and view them. Press TRACE and use the left and right arrow keys to investigate the total number of data values in each bar of a graph and to see which values in each graph are rare.

Standard Deviation Basics: Students often find the calculation of standard deviation of a data set both long and difficult. The calculator can provide shortcuts. Although the biggest shortcut is the 1-variable statistics feature, students often understand a statistical measure such as standard deviation more clearly if they first have experience calculating it with pencil and paper. For this reason you might want to postpone the 1-variable statistics technological shortcut until *Penny Weight Revisited.*

Instead, consider using the calculator's lists as a spreadsheet to merely assist in the calculations of standard deviation, as follows. Enter the original data in list **L1**. Define list **L2** as **L1 – mean (L1)**, which calculates the difference between each data value and the mean, or the *residuals*. Define list **L3** as $(L2)^2$, which squares the residuals. Quit to the Home screen and enter $\sqrt{\dfrac{sum(L3)}{dim(L3)}}$, which calculates the square root of the quotient of the sum of squared residuals and the number (dimension) of data values.

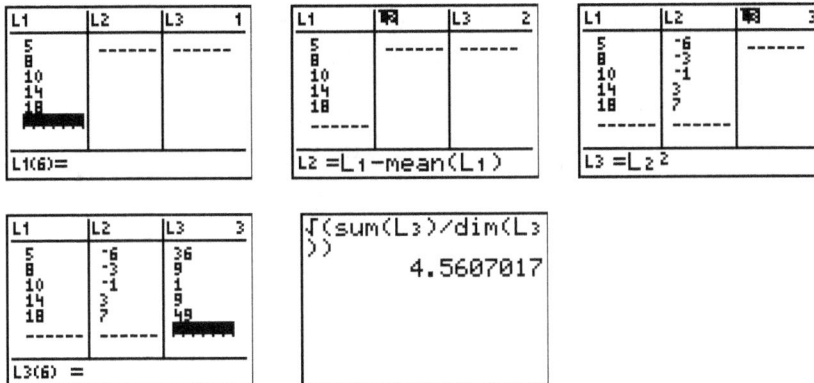

This process requires several 2nd [**LIST**] commands. The commands **mean(** and **sum(** are in the **MATH** submenu, while **dim(** is in the **OPS** submenu. Refer to "Calculating the Mean and Median of a Data Set" in *The Overland Trail Notes for the TI-83/84 Family of Calculators* for a brief review of accessing the **LIST** menus and the **mean(** command. *Note:* While accessing the **MATH** submenu, students may notice the **stdDev(** command. This is actually used to calculate a *sample* standard deviation (with divisor $n - 1$), not a *population* standard deviation (with divisor n), and therefore does not correspond to the calculations described in *Standard Deviation Basics.*

Penny Weight Revisited: The note "Calculating Statistical Measures" demonstrates how to find and use the calculator's 1-variable statistics feature. These instructions show how to find the mean, standard deviation, and sample standard deviation of the sample penny weights listed in *Penny Weight Revisited*. Because students will need to calculate these statistical measures throughout the rest of the unit, you might encourage them to save these notes to use as a reference later.

A Picture Is Worth a Thousand Words: A graphing calculator is a good tool for creating and exploring misleading graphs, because its technology allows you to adjust axes and rescale a graph quickly. You could have students enter one of their graphs from Question 1 as a scatter plot into a calculator with overhead viewing capabilities. For directions on making a scatter plot, see "Finding a Line of Best Fit and Using It for Prediction" from *The Overland Trail Notes for the TI-83/84 Family of Calculators*. Then the entire class can view the graph together and experiment with changing the window and distorting the data set.

Graphing Free-for-All: Along with graphing calculators, you might have copies of "Graphing Basics" from *The Overland Trail Notes for the TI-83/84 Family of Calculators* for reference.

There are four basic ways to describe a function: as a table of values, as a graph, in algebraic terms, and as a representation of an actual situation or event. A graphing calculator can neatly connect the first three of these representations because a it allows you to move easily between a function's table, its graph, and its algebraic equation. The "Graphing Basics" discussion from *The Overland Trail* explains the graphing and table features. The note on "Using Function Notation on a Calculator" from *The Pit and the Pendulum* describes how to use function notation to generate individual outputs. Because function notation can be a difficult concept, students might find experimenting with the calculator's function notation helpful, especially because they can generate function outputs easily on a calculator.

You might need to remind students about using parentheses as they input functions. It's easy to make mistakes when entering a square-root function or ratios. To enter the function $y = \sqrt{3x + 5}$, you need to put parentheses around $(3x + 5)$. To enter $y = -\dfrac{2}{3}x$, you need parentheses around $\left(\dfrac{2}{3}\right)$ or $\left(-\dfrac{2}{3}\right)$.

An Important Function: As students collect and record on a chart the length-versus-time data from their pendulum swings, you might have someone enter the same data into a calculator with overhead viewing capabilities. The curve fitting and predicting over the next few days work well on a graphing calculator. It will be convenient to have the data already entered and ready to use.

Graphs in Search of Equations III: Discuss this activity while a calculator screen is projected. First match the calculator window with the one in the homework and then use **ZSquare** from the **ZOOM** menu so that the graphs don't appear distorted. Now enter different student suggestions for functions. Trace along each function to see if the values match those from the homework. Also, check out a table of values by pressing 2nd [**TABLE**].

The Thirty-Foot Prediction: As students fit a function to their length-versus-period pendulum data and use the function to predict the period of a 30-foot pendulum, they might refer to "Finding a Line of Best Fit and Using It for Prediction" in *The Overland Trail Notes for the TI-83/84 Family of Calculators.* The instructions for a best-fit line are almost exactly the same as those for any other best-fit function.

The lists shown here display a sample data set for the lengths and times of different pendulums. In this particular example students measured a 1-foot, 2-foot, 3-foot, and so on, all the way up to a 7-foot pendulum. The time measurements in list **L2** show the number of seconds elapsed for 12 periods.

L1	L2	L3	2
1	14	------	
2	19		
3	23		
4	26.5		
5	30		
6	33		
7	35.3		

L2(7) =35.3

The graph of the sample data set is shown. Students should notice that the graph follows a nonlinear model, especially if they imagine adding the point (0, 0) to the data set. (The origin makes sense as a data point if you think of a pendulum with length 0 as having a period of 0 seconds.) They should guess a function like $y = 14\sqrt{x}$ as their best-fit function.

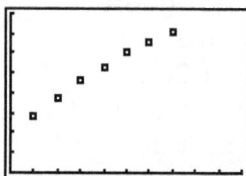

Students should choose this function because they recognize the $y = \sqrt{x}$ shape. They might also look at the table of values and realize that a coefficient of 14 in front of a square-root function makes the ordered pair (1, 14) a solution. This is the first ordered pair in their lists. As it turns out, this function is a rather good fit. This screen shows the data set and function graphed simultaneously.

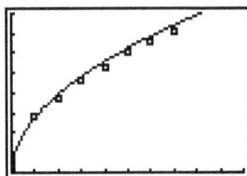

An analysis of the pendulum experiment using physics gives the function $y = 2\pi\sqrt{\dfrac{x}{g}}$

where g is the gravitational constant of approximately 32 feet/second^2. This equation models the relationship between x, the length of a pendulum in feet, and y, the length of a single period in seconds. (This model assumes that the amplitude of each swing is small.) If you simplify this model and modify it to match 12 periods, you get $y = 13.33\sqrt{x}$. (The coefficient 13.33 is rounded.) Notice that this function closely matches the function $y = 14\sqrt{x}$.

The Pit and the Pendulum Portfolio: Students may want to include printouts of calculator screens in their portfolios. To print a calculator screen, you'll need to link the calculator to a computer using the TI Connect™ cable and software. Once students transfer the graph to the computer, they can print from the computer or save image files for use in other computer applications.

Supplemental Activities

Some of *The Pit and the Pendulum*'s Supplemental Activities provide opportunities to introduce additional features of the graphing calculator.

Quartiles and Box Plots: The 1-variable statistics feature includes the values of the minimum, lower quartile, median, upper quartile, and maximum, but students need to scroll down to see them all. Similar to the calculator's ability to make frequency bar graphs and scatter plots, **STAT PLOTS** can be used to make box-and-whiskers plots of one-variable data stored in lists. There are technically two kinds of box plots—with and without outliers; the screens below show the settings and results for the option without outliers. When you trace a box plot, you are given the values of the minimum, lower quartile, median, upper quartile, and maximum.

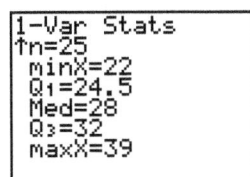

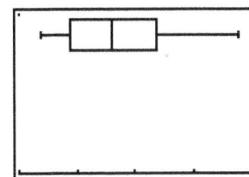

Family of Curves: This activity can be made more interactive with the calculator application *Transformation Graphing*. This free application is distributed by Texas Instruments (visit education.ti.com); it comes preloaded on TI-84 Plus calculators, and it can be downloaded and installed on TI-83 Plus calculators using TI Connect™. (Applications cannot be loaded onto a standard TI-83.) To see if a calculator has *Transformation Graphing* installed, press APPS and look for **Transfrm**. *Transformation Graphing* allows you to define functions with variable coefficients (from the beginning of the alphabet) and then interactively see the effects of changing their values.

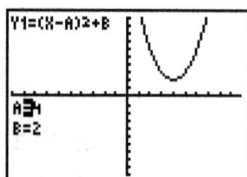

If you choose to use this application, visit education.ti.com and download the *Transformation Graphing Guidebook for TI-83 Plus/TI-84 Plus*, which documents all of the application's features and options.

Making a Frequency Bar Graph

To enter data sets, first press [STAT], highlight **EDIT**, and press [ENTER] to display your lists.

If list **L1** already contains data, place your cursor on the label of the list name and press [CLEAR] [ENTER]. Enter your data set from the activity *Time Is Relative*.

To set up your calculator to build a frequency bar graph, press [2nd] [STAT PLOT]. You should get a screen similar to the one shown here. You will use Plot 1. Press [ENTER] to display the Plot 1 screen showing.

Make your Plot 1 screen match the one shown here by highlighting **On** and the frequency bar graph. Press [2nd] [L1], because this is where you stored your data set. Also, type a frequency of 1.

Check a few things before you press [GRAPH]. First, press [2nd] [STAT PLOT] and make sure all the other plots are turned off. Also, press [Y=] and make sure any functions are either deleted or turned off. You don't want any extra graphs cluttering your screen. Now, press [GRAPH].

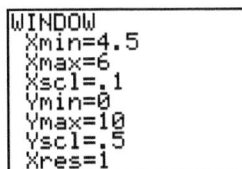

If your window settings have the wrong values, you might get an error message when you try to graph. Even if you don't get an error message, you will probably need to adjust your window to get a better view of your graph. To adjust your window, press [WINDOW]. Select the window values carefully. **Xscl** is an important value because it determines the width of each bar in your graph.

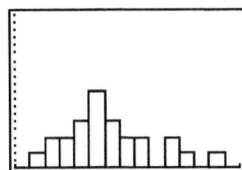

If you have a large data set, a **Ymax** of 10, as shown in the example window here, might be too small to fit some of the taller bars in your graph.

Continued on next page

Reading the Frequency Bar Graph

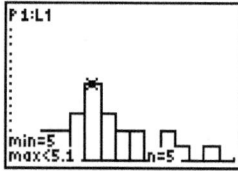

To read values from the frequency graph, press TRACE. Use the left and right arrow keys to move the cursor to different bars. Observe the numbers at the bottom of the screen. In the screen shown here, the cursor is on the tallest bar and the screen shows **min=5** and **max=5.1**. The screen also shows **n=5**. This means that there are a total of five *x*-values greater than or equal to 5.0 and less than 5.1. If an *x*-value occurs on the edge of a bar, the calculator counts that value in the bar on the right.

Calculating Statistical Measures

You can use your calculator to find several statistical measures, including the mean, standard deviation, and sample standard deviation. In this case you will use the data set for penny weights from *Penny Weight Revisited* on page 336. The only variable in this data set is the weight of the penny, which is why these statistical measures are called 1-variable.

Entering Your Data Set

To enter data sets, first press [STAT] [ENTER] to get to your screen of lists. If the list already contains data, put your cursor on the list label and press [CLEAR] [ENTER].

Now enter the weights of the pennies (in milligrams) from *Penny Weight Revisited* into list L1.

Calculating 1-Variable Statistics of Your Data Set

Once you have entered your entire data set, press [2nd] [QUIT] and return to your home screen. Then press [STAT], highlight **CALC** and choose **1-Var Stats**.

Press [ENTER], and the command will appear on your home screen. Enter L1 to ensure that the calculator uses the correct data set for its calculations and then press [ENTER].

You will get a display of different statistical calculations. The display contains so many statistics that you need to scroll up and down to read all the entries. These are the calculations you need to understand for now.

$\bar{x}$ represents the mean of the data set.

S_x represents the sample standard deviation of the data set.

σ_x represents the standard deviation of the data set.

n represents the total number of data items.

Continued on next page

Compare the calculator values for σ and x̄ to the values you calculated for this data set previously. Do these values match your previous calculations?

Recalling a 1-Variable Statistic

Suppose you want to see or use only the standard deviation of the penny weights data set. Press VARS, highlight **Statistics**, and look at the list of statistical measures available in the **XY** submenu.

Highlight σx and press ENTER. The variable should appear on your home screen. Press ENTER again to see the value of the variable. Notice that its value matches your most recent 1-variable statistical calculations.

```
XY Σ EQ TEST PTS
1:n
2:x̄
3:Sx
4:σx
5:σy
6:Sy
7↓σy
```

```
σx
        9.884331035
```

```
Plot1 Plot2 Plot3
\Y1■2X+1
\Y2=
\Y3=
\Y4=
\Y5=
\Y6=
\Y7=
```

```
VARS Y-VARS
1■Function…
2:Parametric…
3:Polar…
4:On/Off…
```

```
FUNCTION
1■Y1
2:Y2
3:Y3
4:Y4
5:Y5
6:Y6
7↓Y7
```

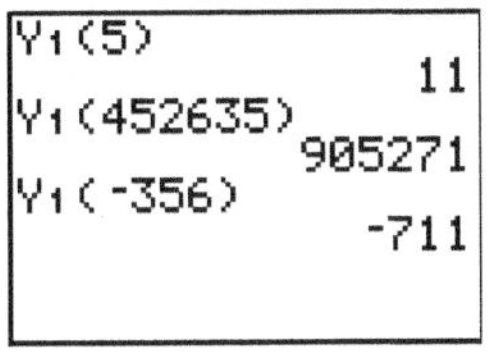

```
Y1(5)
             11
Y1(452635)
         905271
Y1(-356)
           -711
```

Using Function Notation on a Calculator

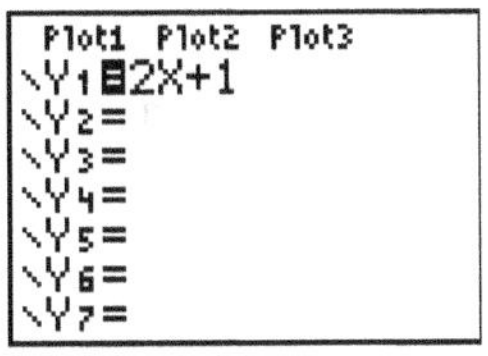

First press [Y=] and enter any function you like for **Y1**. Then press [2nd] [QUIT] to return to your home screen.

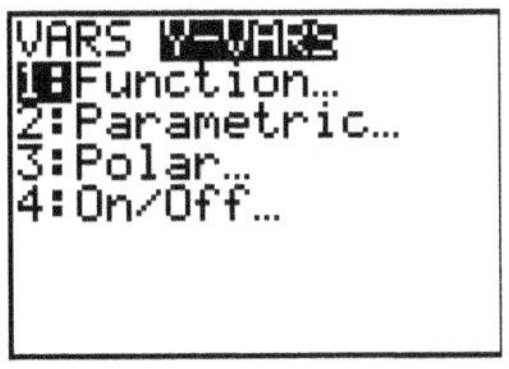

Press [VARS] and then select the **Y-VARS** submenu. Select **Function** and press [ENTER]. Select **Y1** and press [ENTER].

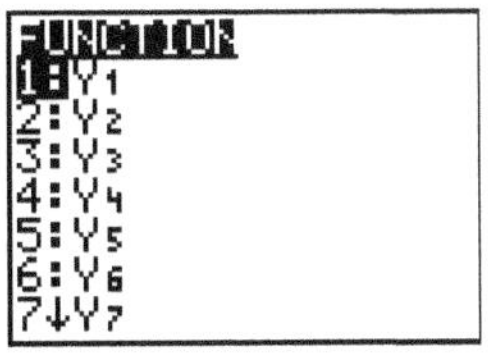

Y1 should appear on your home screen. Use parentheses to mimic function notation, and input any value you want into your **Y1** function as shown here. Press [ENTER] to read the output. Try many different input values. Because accessing the **Y-VARS** menu takes several steps, this is a good place to use [2nd] [ENTRY] and simply edit your last line.